Writing Dancing Together

Also By Valerie A. Briginshaw

DANCE, SPACE AND SUBJECTIVITY

Also By Ramsay Burt

ALIEN BODIES
JUDSON DANCE THEATRE
THE MALE DANCER

Writing Dancing Together

Valerie A. Briginshaw & Ramsay Burt

palgrave
macmillan

First published 2009 by
PALGRAVE MACMILLAN

Palgrave Macmillan in the UK is an imprint of Macmillan Publishers Limited, registered in England, company number 785998, of Houndmills, Basingstoke, Hampshire RG21 6XS.

Palgrave Macmillan in the US is a division of St Martin's Press LLC, 175 Fifth Avenue, New York, NY 10010.

Palgrave Macmillan is the global academic imprint of the above companies and has companies and representatives throughout the world.

Palgrave® and Macmillan® are registered trademarks in the United States, the United Kingdom, Europe and other countries.

ISBN-13: 978–0–230–53564–0 hardback
ISBN-10: 0–230–53564–X hardback

This book is printed on paper suitable for recycling and made from fully managed and sustained forest sources. Logging, pulping and manufacturing processes are expected to conform to the environmental regulations of the country of origin.

A catalogue record for this book is available from the British Library.

A catalog record for this book is available from the Library of Congress.

10 9 8 7 6 5 4 3 2 1
18 17 16 15 14 13 12 11 10 09

Printed and bound in Great Britain by
CPI Antony Rowe, Chippenham and Eastbourne

Contents

Section 4 Two Interviews

Section 5 Insurmountable Memories

Section 6 Reimagining Dancing Together

Section 7 Conclusion

List of Figures

Acknowledgements

We should like to acknowledge and thank our respective universities: Chichester University (Valerie) and De Montfort University (Ramsay) for their support in giving us study leave to enable the completion of this book. Thanks go to the British Academy for a small research grant which enabled us to visit Wuppertal to research in the Pina Bausch Tanztheater archive and to visit Cologne to interview Peter Pabst. Ramsay would like to acknowledge the Arts and Humanities Research Council for funding research leave which allowed him to spend some time in residence with Lea Anderson and her companies. Several of the chapters in the book have arisen from other published work or conference papers and acknowledgements for sources and support for particular essays are provided in the first endnote of the relevant chapter. Certain key people who have been instrumental in our writing relationship are mentioned in the Conclusion to the book. Last but not least we should like to acknowledge and thank all the artists whose work we discuss here, without their stimulating work this book would not have been possible.

Introduction

This book brings together writing from both of us that spans almost two decades. It is a combination of singly authored and jointly written chapters, with jointly written section introductions. Revisiting these chapters and the times in which they were written enables us to explore and investigate the ways in which dance history in particular, and dance studies in general, have changed over this formative period. Significantly, it also enables us to examine the influences and effects of our writing and to re-evaluate the aims we set ourselves at the time in the light of subsequent developments. What is at stake in these pieces is often bound up with where and how our writing is positioned in relation to the dance canon. This book includes two interviews conducted jointly: one with Meredith Monk in 1987, and the other with Peter Pabst, Pina Bausch's scenographer, in 2002 (expanded in 2008), a 'Position Paper' written jointly in 1998, and eleven chapters. The earliest of these is from 1992, the majority have been written since 2000, five by each of us, some of which have not previously been published, and one new joint piece. One of the aims of bringing together the work in this book is to reflect on how and why our shared agendas have formed and evolved.

Certain key themes emerge and circulate through the writing, often reappearing, sometimes in different guises, in different pieces. One of the aims of this book is to track these recurrent themes and reveal the narratives they tell. Key to our writing has always been a concern with situating dance in a social, historical and, importantly, political context and we have often drawn on a range of different methodologies to do this. For example, Valerie's 'Dancing Dicks' chapter, first drafted in 1992, has an explicit feminist agenda to articulate the fetishisation of ballerina as phallus through analysis of choreography, performance and ballet criticism. It draws on feminist and psychoanalytic theories to support its argument. Revealing these signifying elements in ballet and ballet criticism counters and challenges an ahistorical canon, in which classical ballets possess an aesthetic that transcends such meanings, and are seen as timeless masterpieces. 'Dancing Dicks' was very much of its time, although it has never been published. While the insights it offers are of value in themselves, it also serves as a useful reminder that questions posed then are still important and no nearer resolution, but now seem much more difficult to write about.

Feminist political concerns weave their way in different guises through much of the writing: for example, in our questioning of Meredith Monk about perceived female imagery in her work; in the blurring of gender boundaries, which we explore in Lea Anderson's work; and in Ramsay's investigation of how *L'Après Midi d'un Faune* articulates gender ideologies. Indeed, the ways in which we approach feminist political concerns have moved from overt feminist agendas during the 1990s to ones concerned more with gender ideologies and gender blurring in our more recent writing, exemplifying our responses to changing scholarly and socio-political agendas. A recuperation of feminist strategies has perhaps diminished some of their effectiveness without removing or dealing with the problems which feminists have sought to address. Looking back now on the relatively plain speaking of the earlier pieces, we have a sense that there is something valuable in them that is in danger of being lost. We may now appreciate that the issues we were writing about were more complex and fraught with conflicts and contradictions than we then realised. They nevertheless demonstrate a straightforward embrace of commitment that has perhaps subsequently become difficult to sustain within a more conservative and increasingly complicated and fragmented social and political climate. In the 1980s and early 1990s the use of feminist theory heralded an overt concern with deconstructive and poststructuralist theories of identities focusing on issues such as class, race and sexuality. Latterly our writing has been more concerned with a broader political agenda informed by philosophical issues surrounding subjectivity, embodiment and ethics. The political issues of power imbalances associated with particular identities and groups are still there, but the need to find new means in our writing to combat these is an ongoing process.

Our concern to situate and discuss dance in relation to broader social and historical contexts than is normally the case is then overtly political. An interest in power and politics has always informed our writing and weaves its way through the pieces here. In our attempts to re-historicise dance, our writing often challenges the positioning of dance by the canon. For example, Ramsay's chapter exploring Frederick Ashton's *Facade* alongside Lea Anderson's *Elvis Legs,* focusing on the re-appropriation of popular dance forms in the works, challenges the modernist canon which has *de*-historicised and abstracted Ashton's work. This unsettling of the canon is possibly why this chapter of Ramsay's has not been published before now. Ramsay's writing in particular often focuses on situating dances in broader social and historical contexts in order to present a new or different way of seeing dance history. An example is his piece on cosmopolitanism and energetic excess in *Napoli* and *Palermo, Palermo* bringing together dances created 150 years apart, hence its placing in the section of the book we have titled 'Rethinking Temporality: Preposterous Histories'. Much of Ramsay's writing could be included under this umbrella title. He employs similar strategies in his other chapter bringing together the work of Isaac Julien and Meredith Monk. However, these chapters do more than introduce new ways of seeing dance history. They also engage with and introduce new ways of understanding ideas. This tends to be the focus of much of Valerie's writing.

In comparison with Ramsay's essays, Valerie's are less concerned with dance historical trends and positioning *per se* and focus more on readings of individual works alongside philosophical theories. By doing so, she suggests ways of re-assessing or re-figuring positions, perspectives, or ways of seeing things, ideas or ideologies. Although both of us are committed to troubling and problematising the dance history canon, it is Valerie who has often deliberately chosen to expand the range of dance artists normally discussed by dance scholars, and to draw attention to kinds of work and aspects of dance performance that have not received the attention they deserve. In doing so, she has often therefore also needed to develop new methodological approaches. This has led her, over the years, from a theoretical concern with issues concerning representation to a philosophical concern with the embodiment of subjectivity, and the ethical responsibility of which revelations of corporeal vulnerability remind us. In her essay on Bausch's tanztheater, she thus focuses on the corporeality and materiality of the dancers' bodies and explores their vulnerability to demonstrate the potentiality of the reciprocal implications of existing between the body's constructedness and its physicality.

Embodiment and subjectivity are key ideas for both of us. Our investigations of these in and through various dance works are informed by our reading of poststructuralist theory and Deleuzian and post-Heideggerian philosophy. This has led us to focus more on performance and performer–spectator relationships than on choreography, and to explore related topics such as memory and sensation that are bound up with expression and affects, and ethics. Thus Valerie's chapter exploring Emilyn Claid's *Remember to Forget,* alongside Deleuze's discussion of Francis Bacon's paintings, focuses on ideas about the violence of sensation in both artists' work that open up some of the workings of memory.

Sensation also figures in a different guise, that of affective qualities, in Valerie's exploration of Jonathan Burrows's and Matteo Fargion's *Both Sitting Duet* informed by Deleuze's theories of difference and repetition, where she suggests seeing the dance in terms of relations between self and other that matter, introducing an ethical concern. Ethics also figures in Ramsay's chapter on recent works by Yvonne Rainer and Xavier Le Roy, through his use of Spinoza's philosophy of sociability which concerns a flow of affects or sensations between individuals.

The circulation of ideas within and between these various pieces of writing should be evident. By assembling and juxtaposing these pieces and through our introductions to them, our intention is to reveal the many interconnected threads of ideas and themes that crisscross our writing and the stories they tell, and the ways in which these have changed over the years. These interconnections are perhaps not surprising. Over the years we have attended many dance workshops, classes, festivals, performances and conferences together, discussed them avidly, and importantly, read each other's writing closely and critically, and discussed the ideas that have inspired and informed it. Our writing relationship, writing dancing together, has resulted in us reading dance criticism against the grain, [re]writing preposterous histories, uncovering uncontrollable intensities, rediscovering insurmountable memories, and ultimately, re-imagining dancing together. Each of these, to a certain extent, represents a political strategy we have employed at a particular time to address pressing issues. We are well aware that we have not always succeeded in pushing ideas forward as far as we hoped to do, and we recognise the continual need to reassess and rework ideas and rethink strategies for addressing and articulating timely and relevant issues and concerns. This book, in its revelation of the writing dancing journeys we've taken together, maps our changing strategies against the changing social, historical and political times in which we've been writing. Through revealing and reflecting upon the underlying forces that have contributed to the development of our ideological and aesthetic agendas, we hope to point towards ways in which we and others who share similar concerns can continue negotiating them in the future.

Section 1

Against the Grain: Sexuality and Ballet Criticism

Introduction

The two chapters in this section, which respectively look at mid-nineteenth-century ballet and the work of the Ballets Russes, examine the discursive operations of eroticism and male sexuality. By reading the critical reception of the ballerina and Nijinsky's *L'Après-midi d'un Faune* against the grain, we reveal the operations of normative discourses of sexuality through which contemporaries interpreted these ballets. Eroticism was an inadmissible sub-text in Western European nineteenth-century critical discourse as a consequence of the workings of historically specific ideologies of gendered behaviour and propriety. It remained hidden in order to tease. It wouldn't have been sufficiently titillating and the critics' writing wouldn't have been so effective if the eroticism had been openly expressed. Whereas the first chapter reveals the hidden workings of normative male heterosexual desire, the second explores the moment when a modernist intervention disrupted these discursive operations. This was in part because Nijinsky, as a Russian, had not been conditioned to conform with Western European norms of propriety; his modernist reworking of ballet thus made explicit what had been implicit in nineteenth-century French literature, and by doing so caused a scandal.

In some ways 'Dancing Dicks', originally written between 1992 and 1994, may now seem old fashioned. It is included here in order to fill a gap in recent dance scholarship. Feminist re-workings of Lacanian psychoanalytic theory haunted dance scholarship during the late 1980s and early 1990s, although there were no thoroughly worked through examples of it published in relation to dance. Another reason for publishing this now is because during this 15-year period, there has been a backlash against feminist writing in general (see Faludi 1991, 1992), exemplified in dance studies in particular in the work of writers such as Banes (1998), and Carter (1999, 2001).

In our opinion, it is essential to acknowledge the difficulties associated with an approach to feminist scholarship that, by being single-mindedly ideological, fails to take into account the complexities of women's actual lived situations. During the period between the writing of these two chapters, we have had to negotiate our way through a minefield of ideologies. Whilst in 'Dancing Dicks' Valerie's approach drew on psychoanalytic theories, Ramsay's writing on *L'Après-midi d'un Faune* involved the use of methodologies informed by the work of Michel Foucault. Psychoanalytic theories are concerned with the repression of symptoms. Foucault uncovers processes which hide thoughts in order to enable them to be rediscovered, so that such rediscoveries can be used to exert power over subjects. This shift from psychoanalytic to Foucauldian theories exemplifies ways in which we have, in the course of our writing, attempted to address the complexities of lived experience without abandoning the political. The first chapter is informed by theories that posit desire for the other as predicated on lack and thus closure, whereas the second is informed by approaches that see desire as open-ended and productive. These latter concerns inform much of the rest of this book.

1
Dancing Dicks – A Case In Point(e)[1]

Valerie A. Briginshaw

Introduction

Nearly thirty years ago an article entitled 'Alas, Alack the representation of the ballerina' by the British performance artist, Rose English, was published in a special women's issue of *New Dance* magazine. When I first wrote this chapter in 1994, her radical (in 1980) claim that the fetishism of woman as phallus is recognisable in much ballet, particularly in the *pas de deux*, had not been explored by any other dance writers, so I decided to investigate it further.[2] In 1996, Susan Foster's 'The Ballerina's Phallic Pointe' explored the gendered nature of ballet originating in the Romantic period claiming that the ballerina is the phallus and the *danseur* 'embodies the forces that pursue, guide and manipulate it' (1996: 3). Despite our ignorance at the time of each other's work, Foster and I were making very similar assertions. There are, however, distinctive differences in our accounts which render them complementary. Whereas I investigate the fetishisation of the ballerina as phallus in some depth, examining conflations and distinctions between penis and phallus, fetishism, the roles of desire and fantasy, specularisation, the gaze and the climax, Foster's essay situates some of these ideas and claims within a broader historical account of nineteenth-century Romantic ballet. My account shows how the ballerina is fetishised as phallus in a dance example – the *Giselle* Act 2 *pas de deux* – and in some critics' writing – Theophile Gautier's contemporary descriptions of nineteenth-century Romantic ballets and Robert Greskovic's contemporary account of a re-contextualised *pas de deux* from Karole Armitage's *Watteau Duet* (1985).

Rose English's claims

Rose English situates her discussion of representations of the ballerina claiming that in classical ballet, what women dancers represent on stage 'is determined by the dominant patriarchal ideology in which the works are and were conceived, written, presented and seen' (1980: 18). Recounting that nineteenth-century ballerinas were 'feted and sought after by male ruling class admirers,' she asks why the men were 'driven to such distraction by the women on stage?' (ibid.). Drawing on feminist applications of psychoanalytic theory, particularly the work of Laura Mulvey (1975), she claims that ballerinas are fetishised as phallic in the sense that fetishism, according to Freud, is a phallic replacement. The ballerina is eroticised. Phallic meanings are given to parts of her body or all of her. English illustrates her claim with a detailed and convincing graphic description, which it is worth quoting at length, since I explore these ideas further:

> [The female dancer's] leg, on point, is continuously stiff. Her feet end in firm pink points from which her body rises [...] the action of pointing the foot results in a flexed calf which swells and hardens [...] the muscles of the thigh and buttock also expand and firm [...] the leg on point presents a long stiff shape. A *pas de deux* between a man and a woman involves the man lifting the woman [...] The ballerina rises from the man's waist, from his crotch [...] He carries her erect [...] her legs remain stiff [...] He handles her as he would his own penis. Fondly he holds the phallus in his arms, longingly he looks into his princess' eyes, ecstatically he lifts her. His hands around her long, stiff tube of a body. Easily he holds and moves with her. Flying, she is his own.
>
> (English 1980: 18)

English shows how the ballet narratives often contribute to this fantasy by describing the opening scene of *La Sylphide* (1832), where James, the hero, is dreaming in a chair by the fireside. His hands are resting on his lap, in the fur of his sporran. The sylph is at his side. English comments: 'A more overt illustration of male masturbatory reverie I cannot imagine' (ibid.). Having described James's search for the sylph and the *pas de deux* where 'he carries and wields her about with increasing fervour until finally she expires', English claims, 'Her death, the point when she at last goes limp' is 'the orgasm of the phallus that she represents in the fantasy of the hero' (ibid.: 19). English asserts this happens repeatedly in ballets.

The role and value of psychoanalytic theory

Feminist applications of Freudian and Lacanian psychoanalytic theory can inform and critique the content and reading of representations in cultural texts such as ballets, because the central discovery of psychoanalysis – the unconscious and the workings of desire and fantasy within it – can be seen to contribute to the production and consumption of such texts. Psychoanalytic theory can illuminate the ways in which representation captures audiences and draws them into its field and effect by the management of pleasure, fantasy and desire. According to Freud, unconscious fantasies are represented in art, and psychoanalysis has revealed that sexual identity 'is produced through the agency of fantasy' (Burgin 1992: 87). Following this, these fantasies, which contribute to the formation of sexual identity, can reinforce and perpetuate patriarchal power relations founded on constructions of sexual identity.

Penis/phallus distinction

The terms 'penis' and 'phallus' are conflated in English's account as in much psychoanalytic theory. Judith Butler argues that Freud himself may have been proposing a conflation of the two; if this is so, she suggests, the male genitals 'function in a double way' and become 'the site of a textual vacillation [...] between penis and phallus' (1993: 61). The difference between penis and phallus in Lacan's writings is spelt out clearly by Kaja Silverman, who explains:

> Although the phallus is naturalized through its imaginary alignment with the penis, it is never more or less than a distillate of the positive values at the center of a historically circumscribed symbolic order. Those values stubbornly resist embodiment, designating a grammatical rather than an existential position – the place occupied by the name rather than the body of the father. Each of the dominant discourses which make up the larger symbolic order helps to define and localize the phallus by 'imagining' or 'fantasizing' a speaking subject – a subject authorized to command that discourse's power-knowledge. The phallus is in effect the sum of all such speaking subjects.
>
> (1988: 30)

Whereas the penis is part of the male anatomy – a material, sexual object – the phallus is a symbol. The origins of the term 'phallus' come from classical antiquity where it was seen as 'a representation of the erect male organ and a symbol of sovereign power' (Macey 1992: 318).

There is an important link here with ballet, whose technique, based on the philosophical principles of classical harmony, order, hierarchy and authority, has been passed down from the French Renaissance courts, directly to the Paris Opéra, home of Romantic ballets such as *La Sylphide*. The 'long stiff shape' of the ballerina's leg on pointe and the 'long, stiff tube of a body', that English describes, are manifestations of the important classical 'sense of line' in the body. This is a central element of the ballet aesthetic, which was extended and developed with the introduction of pointework in the Romantic period. This hard, strong, upright or erect shape of the ballerina is a visible sign of the discipline of classical ballet technique. It is achieved through the ordering and institutionalised authority of the technical code, principles or laws of the art. Silverman, in her Lacanian discussion of the phallus, stresses its role as a distillate of the positive values at the centre of a historically circumscribed symbolic order and its role as the sum of the speaking subjects authorised to command the power-knowledge of dominant discourses. In other words, the phallus represents the law of the father (in psychoanalytic terms) or of patriarchy. Classical ballet technique exemplifies one of the dominant discourses to which Silverman refers. Psychoanalytic theory, when applied, can thus reveal that ballet's involvement with patriarchal power operates on an unconscious level through the domination of the phallus in fantasy. The image of the ballerina is often read as, or functions as, a reminder or representation of the phallus or the law of patriarchy.

The implications of such readings are particularly evident in the *pas de deux*, where the ballerina is supported and lifted by her male partner. Here sexual possession, ownership, control, display and pride all become part of the fantasy, reinforcing a male heterosexual view that mediates and makes difficult the possibilities of any other readings from female heterosexual or male and female homosexual positions. The female is defined as heterosexual from a male viewpoint, masking any possibility of her own sexuality or desire.

My focus is on the fetishisation of the ballerina as phallus. 'Dancing Dicks', associated with the penile imagery of the ballerina cited by Rose English, exemplifies the textual vacillation between penis and phallus that Butler refers to in Freud's writing and I argue is also evident in subsequent readings of discourses such as ballet. This title is not intended to suggest that the ballerinas themselves were or are 'dancing dicks'. On the contrary, as I show, they often have more power than this derogatory term implies and they also, in more contemporary postmodern works, subvert some of the traditional imagery and associations of the ballerina on pointe. 'Dancing Dicks' is rather the result of one way of reading the ballets, which has often dominated in the past and is still evident.

Fetishism

Fetishism, for Freud, occurs when 'the normal sexual object is replaced by another which bears some relation to it' (1962: 19). Freud cites the foot as 'an age old sexual symbol which occurs [...] in mythology' (ibid.: 21) and he suggests that an explanation for this fetishistic preference for the foot is 'found among the sexual theories of children [where] the foot represents a woman's penis, the absence of which is deeply felt' (ibid.). In Freud's account the female figure is problematic for the male viewer because her lack of a penis is an unpleasant reminder of his first recognition of this lack, and the fear of castration that it evokes. This fear of castration can be seen as fear of a loss of power both in material, sexual terms with the loss of the penis, and in phallic, symbolic terms as the loss of patriarchal authority, hence the *desire* to replace the lost object with a fetish. Freud claims that several cases of foot fetishism were derived from the scopophilic instinct (the pleasure in looking and, importantly for Freud, one of the drives of sexual development) which, when seeking to view the woman's genitals from below, was halted 'in its pathway by prohibition and repression' (ibid.). So the child substituted the woman's genitals which in childhood, according to Freud, were imagined as male, with the nearest thing that resembled them, the shoe or foot and, although Freud does not mention this, the fetish when looking up from below could surely be extended to include the legs.[3]

Ballerinas' legs have certainly always proved a fascination for male audience members. A 1798 cartoon of the Lord Chamberlain and Bishop of Durham looking up the legs of a dancer supposedly to check that the length of her skirt was not immoral (see Figure 1) and the following description of a ballet audience by a social observer in 1843, illustrate the way in which men were 'driven to distraction' by the legs of the ballerinas:

> We perfectly recollect admiring the emotion of several ancient aristocrats in the stalls, on the recent appearance of the legs of Fanny Elssler. We thought that we observed one aged and respectable virtuoso shedding tears; another fainted in his satin breeks and diamond buckles; one appeared to go mad, and bit his neighbour's pig-tail in half in sheer ecstasy. Oh! the legs of Fanny displaced a vast deal of propriety, and frightened sober men from their prescribed complacency.
>
> (quoted in Guest 1972: 63)[4]

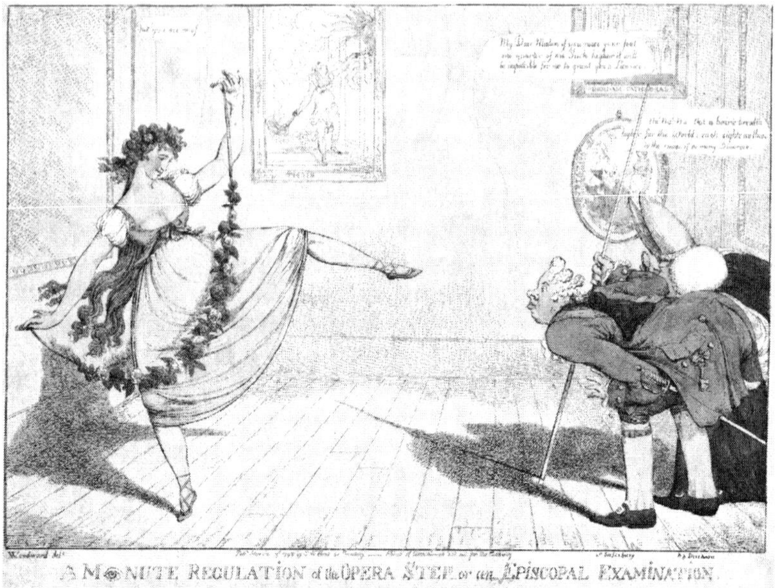

Figure 1 'A M[rebus eye]nute Regulation of the Opera Step, or an Episcopal Examination'

Courtesy of the British Museum.

Freud also mentions the shoe or slipper as a fetish object and he claims that psychoanalysis has shown the importance of a coprophilic pleasure in smelling associated with the feet and shoes as fetishes. He cites the importance of 'dirty and evil-smelling feet that become sexual objects' (1962: 21 footnote). The dirty and smelling characteristics could presumably also be extended to apply to shoes. Ballerinas' used ballet shoes, inevitably dirty and smelly, have been treasured trophies and souvenirs since the early 1800s when pointework first developed. The father of the nineteenth-century ballerina, Fanny Cerrito, who was one of her greatest fans, apparently always carried one of her old ballet shoes around in his pocket.[5] Russian balletomanes allegedly cooked the ballet shoes of Marie Taglioni in a sauce to eat at a banquet! A photo-spread devoted to Margot Fonteyn, in a British newspaper's colour supplement indicates that ballet slippers have survived as treasured trophies into the 1990s. It includes a photo of a discarded pair of her old and dirty ballet shoes and the male writer's comment: 'trophy hunters would have wept at the contents of her dressing room rubbish bin that night' (Money 1992: 18).

Other parallels with ballet are evident in Laura Mulvey's suggestion that Allen Jones's pop art paintings are fetishistic. She claims 'the high heel on high-heeled shoes, a classical fetishist image, is both a phallic extension and a means of discomfort and constriction' (1973: 8). Freud cites Chinese footbinding as fetishism, where he stresses the mutilation suffered by the female foot. Anyone who has danced on pointe or seen the feet of women who have, will attest to the pain and often subsequent mutilation suffered, due to constriction and binding of the feet to maintain balances. Chinese footbinding also suggests the reinforcement of power relations since constriction is a form and symbol of control. As Jung Chang asserts, describing her grandmother's footbinding: 'the sight of a woman teetering on bound feet was supposed to have an erotic effect on men, partly because her vulnerability induced a feeling of protectiveness in the onlooker' (1991: 31).

The nineteenth century was important for the development of fetishism. Naomi Schor comments that the convergence of Freud's theory of psychic fetishism and Marx's theory of commodity fetishism in the nineteenth century underlines the historicity of these concepts which both 'seem bound up with the history of capitalism in its ascending and triumphant phases' (1992: 116). Place is important in these constructs; Schor comments on the widespread evidence of fetishism in 'nineteenth-century *French* representations of femininity' (ibid., my emphasis). The Romantic ballets, where the image of the ballerina on pointe first appeared, were mainly produced in the Paris of the first half of the nineteenth century. The rapidly changing face of the city then, with the increase in bourgeois leisure pursuits designed as profit-making industries, affected ballet. The centre of ballet production, the Paris Opera House, was a commercial institution run on profit-making lines by men for men. Unlike theatre today, most of the audience then was male. Janet Wolff (1990), discussing the separation of public and private spheres in the nineteenth century, argues that the public sphere was a masculine domain. She cites George Sand in 1831 (the year before the first production of *La Sylphide)* dressing as male to be free to enter the public sphere to learn about the ideas and arts of her time, including visiting the theatre and sitting in the pit. Although women were allowed into the theatre boxes or loges, the pit, or stalls, where audience behaviour was raucous and unruly, was inhabited by men. A theatre visit is described in the 1862 'Journal' of the Goncourt Brothers (whose fetishistic writing features later):

> It is wonderful what a centre of debauchery the theatre is. From the stage to the auditorium, from the wings to the stage, and from

> one side of the auditorium to the other, invisible threads criss-cross between dancers' legs, actresses smiles and spectators' opera-glasses, presenting an overall picture of Pleasure, Orgy, and Intrigue. It would be impossible to gather together in a smaller space a greater number of sexual stimulants, of invitations to copulation.
>
> (in Bershad 1992: 101)

Note the emphasis on the dancers' *legs*. It is evident from this contemporary description, as Deborah Bershad comments, that the theatre 'became a setting saturated with sexual meanings' (ibid.: 100).

The ballerinas were created as stars, much like Hollywood film stars. Lithographs, pictures and postcards of them were produced in large numbers. They featured on the covers of sheet music and in popular contemporary periodicals. Souvenirs and mementoes abounded, including porcelain figurines and ceramic cigarette holders representing ballerina's legs, indicating the fetishism that contributed to these performers' reception. The ballerinas' star status is not surprising given the contemporary artist's view of 'woman'. Wolff quotes Baudelaire (1863) who claims in 'The Painter of Modern Life':

> Woman [...] for the artist [...] is far more than just the female of Man. Rather she is a divinity, a star, a glittering conglomeration of all the graces of Nature, condensed into a single being; the object of the keenest admiration and curiosity [...] She is a kind of idol, stupid perhaps, but dazzling and bewitching, who holds wills and destinies suspended on her glance.
>
> (1990: 43)

Baudelaire's words are particularly apt for ballet. His mention of woman's association with 'the graces of Nature', for example, brings to mind many of the female roles in popular contemporary ballets: sylphs, naiads, peris, wilis were all forest or water spirits. Certainly these ethereal beings were 'dazzling and bewitching' idols for the audience and constructed as such by the narrative and form of the ballets. All these features, and the dancers' star status, contribute to the ballerina's fetishisation.

Desire/fantasy and the admiration of virtuosity

The fetishisation of the ballerina as phallus is driven by fantasy and desire in the male unconscious. These masculine fantasies contribute to the construction of gendered subjects in patriarchy. They are part of the learning process of what it means to become masculine or feminine. These processes confirm and help to produce patriarchal power relations through the operation of desire and pleasure. As Juliet MacCannell explains, Lacan has revealed that 'desire fundamentally affects all human relations [...] and skews them in the direction of a single form of relation, domination and servitude' (1992: 65). This is the dominant way of thinking about desire from a psychoanalytic perspective. However, the singular nature of this theory, and the fact that it is based on lack, have been problematised by some poststructuralist theorists, notably Deleuze and Guattari, who posit an alternative notion of desire as multidimensional and productive which works very differently. As Elizabeth Grosz indicates, this has been derived from Spinoza's ideas of understanding the body 'in terms of what it can do [...] the linkages [...] transformations and becomings it undergoes'[6] such that desire can also be seen 'as what produces, what connects' (1994: 165). Whilst I concur with these alternative theories and have used them myself, I still feel that many readings of Western culture are so steeped in a particular way of seeing the world that feminist readings of Freudian and Lacanian psychoanalytic theory can usefully expose some ways in which these discourses work.

Freud considers scopophilia (pleasure in looking/an eroticised look) to be an important drive of sexual development, part of the learning process of how to become masculine or feminine. Grosz claims that scopophilia occupies an 'especially significant place' for Freud 'through its direct links to the desire for mastery [...] and the desire to know' (1992: 447). Classical ballet disciplines bring with them conventionalised presentation modes involving virtuosic execution of classical ballet technique which facilitate the ballerina's submission to the male spectator's 'desire for mastery' and his 'desire to know'. Technical skill and virtuosity are vital contributions to the construction of the ballet spectacle. English's description of the ways in which a ballerina can be seen as a fetish object illustrates the extent to which fetishisation is dependent on the dancer's virtuosity, particularly in the *pas de deux* when she is lifted or supported in a balance by her male partner. Mulvey highlights 'the increasingly insistent theme of women balancing' (1973: 11) as a device for constructing fetishistic images of women in Allen Jones's paintings. She asserts: 'female figures hang suspended, at their peak, at the point of coming down (the phallic reference is obvious)' (ibid.). She argues that Jones, by developing the theme of balance in paintings of women circus performers, equates woman with phallus, illustrating Freud's claim that 'the remarkable phenomenon of erection which constantly occupies the human phantasy, cannot fail to be impressive as an apparent suspension of the laws of gravity' (ibid.). Citing 'dancers on points, waitresses carrying trays, women [...] teetering on high heels', she concludes: all are 'forced to be erect and to thrust vertically upwards' (ibid.). The audience's admiration of this mastery contributes to the fetishisation process and involves specularisation.

Specularisation of woman[7]

Kaja Silverman situates the construction of woman as spectacle in a specific historical context. She suggests that there has been a '"revisualization" of sexual difference' or a 'shift in the general terms of cultural reference which occurred between 1750 and the present' concerning 'the insistent equation of woman with spectacle and man with vision' (1988: 24). She claims this shift is evident in 'nineteenth-century discourses from painting and the novel to psychoanalysis and photography' (ibid.) and I would add, ballet. Prior to the mid-seventeenth century, ballet was dominated by male dancers. Female performers began to dominate in the Romantic period. The 'revisualization' of sexual difference, resulting in the despecularisation of the male subject and the hyperspecularisation of the female subject, provided the context for fetishisation of the ballerina. The construction of the ballerina as spectacle and fetish is inextricably bound up with the notion of the male or masculine gaze; the structure whereby woman is seen from a masculine perspective or through masculine eyes (Mulvey 1975). There are at least two ways in which the masculine gaze can be seen in operation in ballet: within the conventions of the *pas de deux*; and within the overall ballet narrative.

The gaze

In the pas de deux, frequently the *danseur* looks at his partner. What may not be apparent, however, is the way in which these looks are coded into the conventions of the form. The ballet dancer, Anton Dolin, says of the *danseur*, 'he is there to focus attention upon her [the ballerina]' (1969: 12), and he 'should keep his whole attention on her and with obvious admiration at what she has accomplished' (ibid.: 15). The eminent ballet writer Arnold Haskell confirms this for male partnering, claiming: 'gallant partnering is rarely noticed when it succeeds, because its very essence is to centre attention on the ballerina' (1989: 7). Ramsay, in *The Male Dancer*, points out that, by doing this, the *danseur* is negotiating the audience's gaze: by 'gazing at the ballerina, the male partner identifies himself with the males in the audience and is in turn available to be the bearer of their looks' (Burt 1995: 155). Thus the audience's gaze is focused and directed by the *danseur*, shown what to look at and how to look, with admiration, at the spectacle of the ballerina on display. Although these admiring looks of the *danseur* appear to be a natural part of the ballet narrative demonstrating the hero's love for his partner, if this is so, why does his lover not return these looks and gaze lovingly back at him? In the Act II *pas de deux* from *Giselle* (1841), for example, Albrecht gazes at Giselle almost constantly, whereas Giselle only returns his gaze very occasionally; mostly she is looking out to the audience, reinforcing her display by her partner as spectacle.

Often ballet narratives include the *danseur* watching the female dancing, frequently when she appears unaware of him, suggesting an element of voyeurism. This occurs at the beginning of *Giselle*. Although these moments contribute to the narrative, they also interrupt it because the *danseur's* gaze focuses on the ballerina, directing the audience's gaze towards pure spectacle, which does not further the story. The object of the masculine gaze, because of its fetishistic powers, according to Mulvey, freezes the narrative. Referring to Hollywood movies, she claims:

> the female image as a castration threat constantly endangers the unity of the diegesis and bursts through the world of illusion as an intrusive, static, one-dimentional [*sic*] fetish [...] the fact of fetishisation [...] freezes the look, fixates the spectator and prevents him from achieving any distance from the image in front of him.
>
> (1975: 26)

In ballet, the whole *pas de deux*, often termed a divertissement (an entertainment or distraction only loosely connected with the plot), because of its virtuosity and form, which emphasise the display of the female above all else, also freezes the narrative. Within the *pas de deux* there are also many held moments which freeze the flow of movement in still tableaux when the ballerina is posing, balancing, being lifted. In these, she is almost always in an extended, erect posture. The *pas de deux* is often claimed as the pinnacle of ballet skill and spectacle and as the central pivot on which all ballets turn. Hence the importance of romantic heterosexual love (the *pas de deux* is almost always between a man and a woman) in ballet narratives from *Giselle* (1841) to *Romeo and Juliet* (1965), because this is what is writ large in the *pas de deux*, it is what it ultimately displays and is all about. It is taken for granted and expected that this convention represents heterosexual relations.

Giselle Act II *pas de deux*

The *Giselle* Act II *pas de deux* is a typical example that illustrates the importance of virtuosity and other structural and narrative elements in the content and form of the *pas de deux*, which contribute to and facilitate the fetishisation process.[8] The content includes: a series of balances for Giselle, on flat foot, on pointe and turning, some supported by Albrecht; and supported leaps and lifts. The form of the *pas de deux* involves repetition of this content, which gradually increases in difficulty, virtuosity and speed, interspersed with moments when the two dancers part and return to each other. Throughout, the music gradually builds, accelerating rapidly for the pair's solo variations. These include phrases of jumps, first alone, then Giselle supported by Albrecht, then lifts, both increasing in speed. These include Giselle lifted vertically and held by Albrecht in front of him. The duet culminates in one big lift, a final series of jumps for Giselle at double speed, and a pirouette ending in a balance on pointe. The finale of the whole episode, using very climactic thrusting music, is a series of *grande jetés* performed by Albrecht.

Virtuosity and held or frozen moments are evident in balances, lifts and virtuoso jumps; a series of increasingly difficult technical feats. These moments can be read on at least three different levels: the sheer pleasure of the aesthetic evident in the virtuosity and skill; the fetishisation of ballerina as erect and thrusting penis, particularly in the jumps; and the fetishisation of the ballerina as phallus in the sense of the symbolic power evident particularly in held balances. This series of technical feats has a pattern and rhythm which builds to a climax, emphasising the power and pleasure of the ballerina image. There is an ebb and flow to the form provided by the phrasing of the dance and music, and the meeting and parting of Giselle and Albrecht. This rhythm is delayed and drawn out, extending the pleasure. The insistence of the form is also emphasised by repetition of steps and movements. The gradual building in the form is created by acceleration in the dance and music, an increasing difficulty of the steps, higher lifts, and increasing intensity of the relationship of the lovers shown by a mounting closeness of their two bodies, eye contact and facial expression. These building repetitive patterns and rhythms can be read on the level of the formal aesthetic and on the level of fetishisation of the ballerina because of the connotations with sexual climaxing. *Pas de deux* in general in ballets, because of their placing and role in the narratives, as well as their virtuosity, are regarded as climaxes or high points, focusing attention on the star of the work – the ballerina. The *pas de deux* from *Giselle* is at the very end of the ballet when Giselle has died and become a wili. It is Giselle and Albrecht's last love dance together before their final farewell.

The climax

The *pas de deux* provides a climax in more than one sense. It is an aesthetic and narrative climax within the ballet and it can also be interpreted as a physical or sexual climax or, as English indicates, a 'metaphorical ejaculation', a term used by Susan McClary to account for a similar occurrence in music. Describing an example of the 'climax-principle' in music, which she claims underwrites much eighteenth-, nineteenth-, and twentieth-century music and literature, she argues:

> the piece had thrust [...] A kind of pitch ceiling consolidates, against which melodic motives begin to push as though against a palpable obstacle. As frustration mounts the urgency of the motivic salvos increases; they move in shorter and shorter time-spans, until they succeed finally in bursting through the barrier with a spasm of ejaculatory release.
>
> (1991: 14)

She continues: 'the gesture is only slightly more graphic in pornographic films, in which key structural moments of tension and release likewise are conventionally embodied through what is called the 'money-shot' – close-up footage of a discharging penis' (ibid.).

There are many similarities in terms of the 'key structural moments' that are 'conventionally embodied' in these different cultural products (ballet, music and porn films) to produce climaxes which, in ballet I am arguing, contextualise the fetish of the ballerina. These moments reinforce the fetish through sexual satisfaction. Richard Dyer (1985) describes the important structuring role the narrative elements play in gay male porn films to create the climax. There are similarities in the *pas de deux* regarding structural moments or narrative elements which actively aid the fetishisation process. The two dancers enter, establish contact initially through looking and then through the male touching, supporting, holding, lifting and carrying the female. Throughout, 'foreplay' is evident in their dancing together and apart, circling around each other as they approach in the *Giselle* duet anticipating what is to come. The high lifts, requiring considerable force, energy and thrust and a tense erect body, usually accompanied by powerful thrusting music, provide the peak of the climax, the held moments of ecstasy for the fetishist. These are often followed by solo variations, providing the parting associated with detumescence, or a series of bending, curving, arching movements of the female's torso still in contact with the male, as in *Swan Lake,* presenting a post-ejaculatory wilting of the penile fetish.

Porn film, music, and ballet all use timing or delaying tactics to emphasise the climax, heighten the desire, establish the need, tease the audience. McClary describes this process in music:

> there were images of desire – desire for the satisfaction of what is experienced as an intolerable lack. The principal innovation of seventeenth-century tonality is its ability to instil in the listener an intense longing for a given event: the cadence. It organises time by creating an artificial need ([...] there is no reason one should crave [...] the pitch D; yet by making it the withheld object of musical desire, a good piece of tonal music can [...] dictate one's very breathing). After that need is established [...] tonal procedures strive to postpone gratification of that need until finally delivering the pay-off in what is technically called the 'climax'.
>
> (1991: 19)

The similarities in porn film, music and ballet are clear, notwithstanding the key difference that porn film has as its declared intention sexual arousal, which is not normally an explicit concern of literature, music and ballet. Their persuasive powers are hidden and embedded in the institutional structures that constitute them as discourses. McClary suggests this is where their dangers and power lie and, because of this, the reading is phallic rather than penile. She claims:

> in most post-Renaissance Western music and in virtually all its critical literature, the climax-principle (like the phallus of the classical Greek column) has been transcendentalised to the status of a value-free universal form [...] it is rarely even viewed as sexual [...] it is simply the way music is supposed to go.
>
> (1991: 21)

Similarly the image of a poised or lifted erect ballerina is so familiar that is it seen as simply the way ballet 'is supposed to go'. However, the mechanisms and structures inherent in these cultural forms (their styles, conventions, vocabulary of expression) are reinforcing and producing patriarchal power relations. As McClary argues:

> the omnipresence of this formal pattern in literature and music is part of a larger cultural tendency to organise sexuality in terms of the phallus [...] to impose and maintain a hierarchy of power based on gender [...] [We inhabit] a world rife with phallic posturing [...] literature and music do not simply reflect the world; they help to create and transmit it by reinforcing as pleasurable [...] these habits of thought.
>
> (1991: 19–20)

As I have argued, ballet operates in a similarly fetishistic manner; however, at times it is doubly fetishistic because the writing of the critics can also be fetishistic in the way McClary claims for literature.

Critics' fetishistic writing

Emily Apter characterises the fetishistic writing style of authors, such as the Goncourt brothers, as elaborating:

> an ideal of feminine beauty, charm, and even intelligence matched by a prose style at once frothy and rhetorically fetishistic [...] the expression *rhetorical fetishism* refers to the taste for epithet, mannered syntax, and tropes of hyperbole and accumulation commonly used [...] to render the codes of *féminilité* [...] in the language of fetishisation [...] the verbal substitute for the phallic referent is reified to the point where its origin is forgotten.
>
> (1991: 68–9, her emphasis)

She claims 'the Goncourts complemented this rhetorical fetishism with a more clinical fetishisation of the female body typically found in the work of eighteenth-century authors.' (ibid.: 69). In their focus on, and obsession with, parts of the body and clothing, she claims that writers such as Theophile Gautier, the nineteenth-century fashion writer, ballet critic and scenarist, 'built up a language of libidinously charged vestimentary details' (ibid.: 71).

There is much evidence of such fetishistic writing in Gautier's ballet criticism. When writing of Fanny Elssler's performance of her *Cachucha* in the ballet *Le Diable Boiteux* (1836), for example, he describes her costume and its fit, in precise and intimate details mentioning 'her pink satin *basquine* trimmed with wide flounces of black lace', 'her skirt, weighted at the hem' which 'fits tightly over the hips' and 'her leg, smooth as marble' which 'gleams through the frail mesh of her silk stocking' (Gautier: 1932: 15). This writing could certainly be described as possessing 'libidinously charged vestimentary details'. He continues: 'her slender waist boldly arches and causes the diamond ornament on her bodice to glitter; and her little foot at rest seems to await the signal of the music' (ibid.), teasing his readers with subtle signs of movement such as the glittering brooch, and suggestions of deferment in Elssler's waiting 'little foot'. He uses a string of epithets to describe Elssler's body parts: 'her lustrous eyes', 'sparkling smile!' and 'rosy fingers' (ibid.) and employing extravagant, hyperbolic language with an accumulation of exclamations, he creates a climax: 'With her hand she seems to shake down great clusters of rhythm. How she twists, how she bends! What fire! What voluptuousness! What precision!' (ibid.). His language next suggests a post-ejaculatory wilting: 'Her swooning arms [...] drooping head, her body curves backwards, her white shoulders almost graze the ground' (ibid.). Then, addressing his readers directly: 'Would you not say that in that hand [...] she gathers up all the desires and all the enthusiasm of the spectators?' (ibid.), he suggests that Elssler's dancing is satisfying the audience's longings and, like the *danseur* in *pas de deux*, Gautier negotiates for his readers his male viewpoint or gaze.

In another of Gautier's descriptions of Elssler, his focus, using epithets and making comparisons with ideals of classical beauty, is almost entirely on her legs: 'she has [...] slim ankles; her legs, elegant and well-turned, recall the slender but muscular legs of Diana, the virgin huntress; the knee-caps are well-defined, stand out in relief, and', employing hyperbole, 'make the whole knee beyond reproach' (ibid.: 21).

He then comments at length that Elssler's 'legs differ considerably from the usual dancers' legs,' which he criticises for being over-muscular, and arousing 'the enthusiasm of the old roués in the stalls'. In comparison he claims her legs are 'beautiful', 'like those of an antique statue'. He craves an apology for his preoccupation with the dancer's legs suggesting: 'We shall be pardoned, we hope, for having discoursed at such length on legs, but we are speaking of a dancer' (ibid.), thus associating dance with legs only. Gautier, in fact, called ballet the 'literature of the legs'.

This fetishistic style of writing was very fashionable in the nineteenth century; however, traces remain in some relatively contemporary ballet criticism. In 1985, Robert Greskovic, an American critic, devoted a lengthy article (in *Ballet Review*) to the dancing of Karole Armitage, a classically trained postmodern choreographer, who puts her pointework to very different uses than the Romantic ballerinas. Armitage deliberately tries to subvert classicism and 'make it new' (Salle 2006–7: 10). She said of her work in the 1980s that she was attempting to discover 'a classicism for our time' (in Zimmer 1986: 40) and that she was developing 'a dance style that showed impetuous, spontaneous, rock-oriented urban feelings, in conflict with more disciplined, rigorous, traditionally classical self-presentation' (ibid.).

Greskovic, whose article is revealingly subtitled 'the shoes of the ballerina', focuses his writing on *Watteau Duet* (1985), which Armitage claimed was using 'the tradition of *pas de deux* clichés, which I could juxtapose or turn upside down. It's chivalrous, romantic, psychological, erotic, playful' (Zimmer 1986). In it, Armitage is subverting the fetishism of woman as phallus by her unusual use of pointework and her exposure of the support structure her partner provides. Greskovic might have been parodying fetishistic writing in his article, but having met him in the early 1990s, and discussed these issues, I very much doubt it. He sets the tone by asserting at the outset: 'ballerinas continue to stand "in our minds" on toe-tip, ready and eager to ascend from the ground' (1985: 73). Focusing on Armitage, he writes:

> Armitage's dancerliness is all blended together into a physique that is long, lean, flexible, and lovely. The narrow lengthiness of her build has a singular graphic force [...] Not only is Armitage built on vividly stretchy lines (her small head and narrow hips keep her lengthiness undistracted), but her proportions are also distinguished by the dancerly dominance of her legs. Armitage stands tall, like a living caryatid – narrow as a column, strong as a post, and arresting as a goddess in repose [...] the separate power of her correspondingly large, strong, narrow feet becomes evident [...] Those mighty yet supple legs of hers shoot out [...] and take her through the air with cool force. Her legs, especially the one thrown forward to lead the jump, reveal hip*thigh*knee*shin*ankle*FOOT focus without ever losing awareness of L*E*G!
>
> (ibid.: 74–6, his emphasis) (see Figure 2)

Note the similarity in style, despite 150 years distance, between Greskovic's writing in the twentieth century and Gautier's in the nineteenth, the details of Armitage's physique given, the obsession with her legs, his use of asterisks and capitals to emphasise every single part of her legs and the effect all this has on him. Greskovic asserts: 'seeing

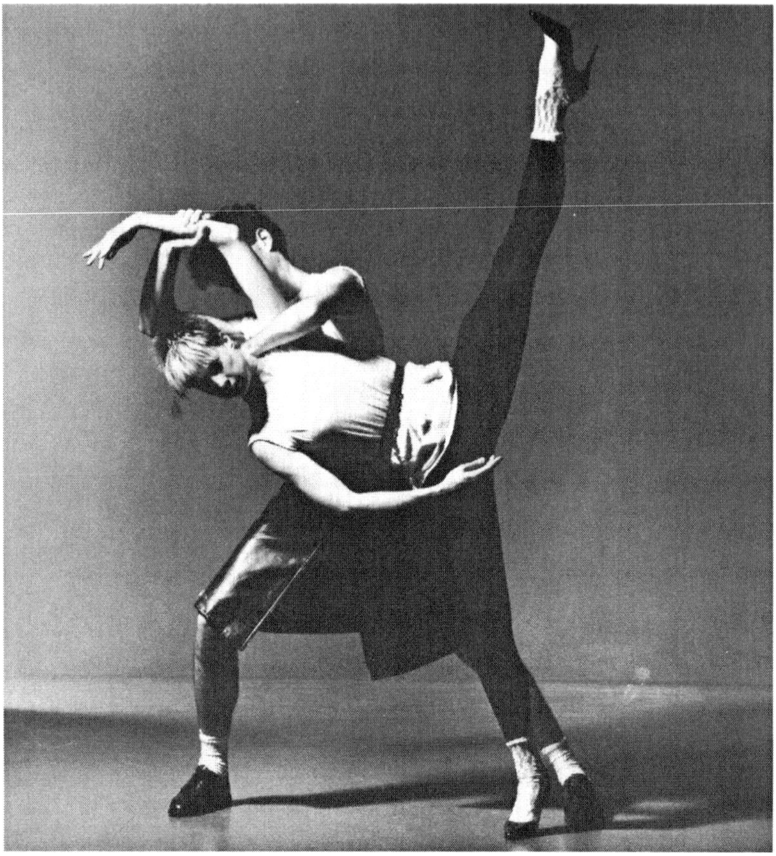

Figure 2 Karole Armitage / Joseph Lennon, *Watteau Duet,* 1985
Photograph © 1985 Lois Greenfield.

[the *Watteau Duet's*] marvels made me hungry for the pieces that likely held other riches' (ibid.: 76) and 'the idea of long, supple, willowy Karole Armitage stretching simple tendus or raising silky battements amid a tornado of sound is as riveting and magical to me as any other *coup de théatre'* (ibid.: 88). Like Gautier and other writers cited by Apter, Greskovic also describes Armitage's costume in detail:

> she wears a crownless helmet that covers her ears and is fitted as well with fingerless gloves, perhaps for archery practice, with laser arrows, in outer space. Her most fantastic detail, however, is on her

> feet – satiny black gaiters that rise to her shins turn her black pointe shoes into sharp pointe boots.
>
> (ibid.: 78)

Note the inclusion of fantasy images and further obsession with the details of Armitage's legs and feet. Describing a later costume, Greskovic provides more evidence of his fetishistic writing with the use of epithets and his detailed focus on her costume and body:

> Armitage displays her formidable legs in further dimension [...] their exemplary power is directly reinforced by the simple tunic-and-tights costume: the little top neatly establishes Armitage's waist, and the black tights can then show her legs in uninterrupted flow from torso to feet – no leotard line separates the top of her leg from the bottom of her torso. The slow, deliberate moves that Armitage makes with her glorious legs remind us of their beauty and power in plain black and white.
>
> (1985: 83–4)

Describing the stiletto-heeled shoes Armitage wears in this section of the duet, Greskovic claims: 'the shiny, sharp, black shoe lends the foot a pretty, high arch and gives it an extra weight, both visual and physical, to complete the leg's shape' (ibid.: 84). The role of high-heeled shoes in fetishistic fantasies has been noted. Greskovic, focusing on Armitage's pointes, claims: 'Armitage reveals something of the actual struggle necessary to keep all the figure's weight on the toetip of the foot' (ibid.: 79–80). The revelation of the 'struggle' brings to mind the importance of balance in fetishisation noted by Mulvey (1973) and also Freud, in his example of Chinese footbinding.

Despite Greskovic's interpretation of Armitage's *pas de deux* as being non-traditional in terms of its power relations, and his comment that she reveals the struggle necessary to maintain her balance, he reinforces his undermining throughout of any subversion in her work, by ending his article with the statement: 'only pedantic, unimaginative thinking could equate classicism with old-fashioned, i.e., not contemporary, aesthetics. Armitage reveals the *joie de vivre* that marks every true classicist' (1985: 89). Greskovic's enthusiasm for classicism can be linked with the classical origins of the term 'phallus'. The rhetorical fetishism in Gautier's and Greskovic's writing is almost entirely phallic with the exception of the first Gautier quotation. Here the rhythm and language ('lustrous', 'quiver', 'fire', 'voluptuousness', 'swooning', 'drooping') can be more obviously read as titillating and sexual. Phallic references in Gautier's and Greskovic's writing, for example, to antique statues (Gautier) and columns (Greskovic) and repeated references to power in Greskovic's writing, abound. Greskovic also reveres the classical line of the body describing Armitage's 'long, lean [...] lengthiness' with its 'singular graphic force' and he appreciates that the costume shows her legs in 'uninterrupted flow from torso to feet'.

Conclusion

The use of Freudian and Lacanian psychoanalytic theory has enabled me to identify some of the previously hidden content of the representations of women in ballet, and to show how these representations or images *work* with audiences to create meanings. The role of desire in this process is evident. It operates as a powerful mechanism of the unconscious, and psychoanalysis shows that the form desire takes is cultural rather than natural, social rather than biological, it shows that it is learned (Gatens 1991). Culture, in this instance ballet, plays a key role in the learning process. Needs are created and satisfied through scopophilic activity, visual pleasure, which is bound up with the masculine look or gaze. At least two gazes are at work – the gaze of the audience and critics, and the gaze of the male *danseur* always focused on the ballerina. Ballet operates in this way as a discourse through which the body is experienced as mediated.

I have examined ballet examples from Paris of the 1830s and 1840s (*La Sylphide* and *Giselle*) and New York of the 1980s (*Watteau Duet*), created by different choreographers, male and female. It may appear that I am seeing the image of the ballerina as a fetish, a 'dancing dick', as fixed, unchanging and universal. I am not. Many factors that contribute to the construction of the ballerina images, are constantly changing – importantly the audiences for these ballets. *But*, as the writing of Greskovic has shown and examples of choreography from different periods and places can corroborate, there are elements of the images (e.g., steps, costume, set, music, story) and of the interpretations of them by some critics which, because of the deep-rooted hold of the classical ballet discourse, appear to have changed little over the last 200 years in the way they operate. The powerful grip that such classical discourses have on the Western world is illustrated by McClary in the case of music. As she suggests, what we need are 'alternate models and images of experiencing pleasure and other, less controlling ways of organising sexuality' (1991: 22).

Not all ballets provide illustrations of this kind of fetishism in action. For example, Fokine's *Le Spectre de la Rose* (1911) opens with a young *woman* sitting in an armchair dreaming of the ball she has attended, clasping a rose she has been given. Her dream is of the spirit of the rose who appears to her in the form of a *male* dancer – the role created by Nijinsky. If Rose English had examined this in her article, what conclusions might she have reached? Is the young woman also experiencing a masturbatory fantasy? What does this do for *female* desire and for the experiences of the audience members – male and female? Although not examined in any detail here, Armitage's representation of the ballerina on pointe exposes many of the conventions of the form, and is material worthy of examination as an alternate model. Does Armitage's work present the ballerina as phallus? How much is she in control of her image? These questions beg others about the ways in which the function of fetishism has changed from the mid-eighteenth century to today, through Romanticism, Modernism and Postmodernism, and about how images of the ballerinas are read by modern, often predominantly female, audiences, but also about the inherent power of the image, which is retained in the fantasies not only of adult audiences but, possibly also, of young girls. Clearly more research is needed to answer these questions. This chapter, in tackling one aspect of this complex field, has, however, revealed how the vocabulary and conventions of classical ballet, given certain circumstances of production and reception, work to reinforce patriarchal power relations by producing 'dancing dicks'.

Notes

1. This chapter would not have been possible without a sabbatical term from West Sussex Institute for Higher Education, UK, in 1992. An early version was given as a Public Lecture in the Center for Theatre Arts, Cornell University, April 1992, where I was Visiting Critic in the Dance Department. Ideas for this came from student and faculty discussions on my *Ballet History* course and from Mandy Merck's *Gay Abandon* course at Cornell University, Spring Semester 1992. The chapter is dedicated to Byron Suber who helped me dream up the title over a late night jug of beer in an Ithaca bar. I would also like to thank Christy Adair, Ann Daly, Janet Sang, Alison Southern, Janet Wolff and Ramsay for reading various drafts and giving helpful comments. A later version was presented as an invited keynote address at the *Body, Gender and Performance* conference University of Stockholm, 2002.
2. Daly (1987–88) discusses the ballerina as spectacle and the object of male desire, but she does not extend this to the notion of the ballerina as phallus in the fetishistic sense.
3. As feminist critics of Freud have pointed out, the castration theory, on which fetishism is based, only really works or makes sense if the child is male.

However, since I am concerned with masculine readings of the ballerina and the role of the male/masculine gaze in these, the theory still stands.

4. It could be countered that male audience members were 'driven to distraction' by the sight of Elssler's legs because prior to this women's legs had remained covered in fashion as well as on stage, and that this solo dance would include little pointework which was still in the early stages of development. The images of the ballerina as phallus that English (1980) describes fit more easily into later Russian *pas de deux* than into the earlier Romantic ballets such as *La Sylphide* (1832). However, my point here is the focus on legs, which are repeatedly mentioned in contemporary critics' writings. The stylistic changes from Romantic to classical ballet through the nineteenth century do not invalidate English's and my claims concerning fetishisation of the ballerina because the principles on which ballet technique is and was based did not change, only emphases on different aspects. In these principles the line of the leg and the upright posture of the body, extended when pointework developed and further extended in lifts, have always been paramount, providing fundamental bases from which different stylistic interpretations and fashions have emerged.
5. Without further research, it is impossible to know *why* Cerrito's father did this and what meanings the ballet shoe held for him: whether it was fetishistic in the Freudian, sexual sense; or whether it was fetishistic in the original sense as an 'inanimate object worshipped by primitive peoples for its supposed inherent magical powers or as being inhabited by a spirit' (Oxford English Dictionary, 1982); a lucky charm of possible spiritual or sentimental value. However, I think it significant that it was a ballet *shoe* that he carried. The shoe may act as a reminder of the foot and the virtuosity of balancing on pointe, rather than as a sexual substitute in the Freudian sense but, as I argue, this kind of admiration of skill and aesthetic idealisation is not innocent. It has important links with operations of desire and fantasy in the unconscious.
6. See also Chapter 11.
7. Specularisation, as well as meaning the creation of spectacle – something to be looked at – also suggests a specific kind of looking. Its root – specular – means 'of the nature of a speculum' (O.E.D.), which is both a surgical instrument for penetrating and examining human orifices and a metal mirror used in a reflecting telescope.
8. This discussion is based on the video recording of *Giselle* performed by the Bavarian State Opera Company with Lynn Seymour and Rudolph Nureyev, 1979.

2

Modernism, Masculinity and Sexuality in Nijinsky's *L'Après-midi d'un faune*[1]

Ramsay Burt

The ballet *L'Après-midi d'un faune,* created by Vaslav Nijinsky and first performed in Paris by Diaghilev's Ballets Russes in 1912, was set to Claude Debussy's *Prélude: L'Après-midi d'un faune* of 1894 that was itself inspired by Stephane Mallarmé's poem of 1876. Mallarmé and Debussy were key figures in the symbolist movement who are recognised as precursors of artistic modernism in the way their works broke through existing conventions to open up what were then new aesthetic and ideological possibilities. Thus Mallarmé's poem initiated radically new, deconstructive poetic practices, while Debussy's *Prélude* instigated a new, post-Wagnerian musical idiom. Although the ballet, through its collaborative enfolding of music, painting, poetry, and dance, continued a symbolist tradition of the total art work, it did so by initiating new modernist modes of choreography and performance. It not only eschewed the vocabulary of classical ballet movement in favour of an innovative, anti-virtuosic movement style, but instituted new ideologies of performative interpretation and new ways of structuring temporality that evoked the experience of modernity. All three works explored highly eroticised subject matter, each in its own time challenging norms of morality and propriety but, crucially, the ballet enacted these through presenting an eroticised male body as the object of its audience's gaze. By doing so, the ballet made disturbingly explicit instabilities and discontinuities within the embodiment of masculine subjectivities that remained implicit but veiled within these earlier works.

Nijinsky's *L'Après-midi d'un faune* (henceforth *Faune*) was one of the more popular items from the Ballets Russes' repertoire and was restaged many times during the twentieth century, resulting inevitably in small changes and modifications of details. Recent attempts to restage the ballet in a form that is closer to the way it was performed in 1912–13 make use of a notated score that Nijinsky produced while under house arrest in Budapest between 1914 and 1915. It was not until the late 1980s that Ann Hutchinson Guest and Claudia Jeschke, working together, succeeded in deciphering Nijinsky's notation system. Using his manuscript score of *Faune*, together with other sources including photographs and drawings of the ballet from 1912–13, they created a new score of *Faune* in Labanotation (Guest and Jeschke 1991). In writing this chapter, I have looked at a video of a production made using this new score together with photographs and drawings, and consulted dancers' reminiscences in order to reimagine the production that critics saw and wrote about in 1912 and 1913. Some viewers and critics welcomed the ballet for its modernist approach, while others found this disturbing. This chapter examines the critical reception of *Faune*, to some extent reading this against the grain, in order to uncover some of the ways in which the dancing body mediated the experience of modernity in 1912. It then considers continuities between the erotic quality of Mallarmé's poetry, Debussy's *Prélude* and Nijinsky's ballet, locating this within an avant-garde, French literary and musical tradition. This provides a framework for discussing Nijinsky's modernist performance of masculine sexuality.

Ballet, modernism and modernisation

What was unsettling about artistic modernism was the loss of the familiar and its replacement with absences that reminded beholders of their own experience of modern living. As Jonathan Crary points out: 'modernisation was not a one-time set of changes but an on-going and perpetually modulating process that would never pause for individual subjectivity to accommodate and "catch up" with it' (2001: 30). Diaghilev in effect challenged audiences and critics to 'catch up' when he and Nijinsky gave an interview six weeks before *Faune*'s premier, during which they described *Faune* as an essay in cubist choreography (Tenroc 1912: 4). This was more than just a desire to appropriate an exciting new critical vocabulary but a reorientation of the Ballets Russes away from the lavish, exotic Orientalism of the productions that initially won over Paris' social and cultural elite groups towards a more avant-garde position. The art historian David Cottington argues that Diaghilev's move towards the avant-garde around 1912 and 1913 played a catalytic role in the acceptance of modernist painting by the Parisian elite. Arguably, after Nijinsky's dismissal, Diaghilev sanctioned no more experiments in avant-garde choreography. Just as some critics found post-impressionist, cubist and futurist painting puzzling and disturbing, the critics who disliked *Faune* were sometimes expressing displeasure that advanced choreography was asking them to shift away from conventional modes of narrative interpretation towards an engagement with theoretical issues.

Hannah Arendt has argued that Walter Benjamin was the first writer to recognise that modernity had created an impassable barrier between past and present. For Benjamin, quotations from the culture of the past always appear to lose their authority. He wrote that, as a collector puts the items of his collection together, he tears them out of the environment in which they had been meaningful but in doing so 'liberates them from the drudgery of usefulness' (1973: 42). A key difference between Mallarmé's poem and Debussy's *Prélude* on the one hand and Nijinsky's ballet on the other lies in the way they quote classical Greek imagery. Mallarmé and Debussy created works at a time when the discourse of classical mythology was still central to cultural discourse. By 1912 new modes of aesthetic appreciation were emerging. Greek vases and sculptures had become part of a wider field of art which included Chinese vases, African masks and sculpture, and American Indian art, all of which were appreciated for their formal qualities in isolation from the cultural meanings they had had within their respective cultural contexts.

This gap between the traditional past and the modern present was at its most evident in the experience of living in the metropolis, and it was as a migrant within the metropolis that Nijinsky, like most of the pioneers of modernist art, lived and worked. Raymond Williams has argued that the experience of exile, and of migrating to a foreign metropolis, was central to the creation of the formal innovations made by the early modernists. At the level of theme, Williams argues, the experience of being an immigrant:

> underlies in an obvious way, the elements of strangeness and distance, indeed of alienation, which so regularly form part of the repertory. But the decisive aesthetic effect is at a deeper level. Liberated or breaking from their national or provincial cultures, placed in quite new relations to those other native languages or visual traditions, encountering meanwhile a novel and dynamic common environment from which many of the older forms were obviously distant, the artists and writers and thinkers of this phase found the only

> community available to them: a community of the medium; of their own practices.
>
> (Williams 1989: 45)

In Nijinsky's case, as I will show, this concentration on the medium itself took the form of a radical reconsideration of the formal language of movement, particularly in relation to time and space. Whereas ballets had up until then matched choreographed steps with music notes in a way that corresponded with an understanding of temporality which derives from classical Greek philosophers, I shall argue that the quality of time in Nijinsky's *Faune* corresponded to new ways of thinking about duration that Henri Bergson had recently proposed. This, I will show, had implications for the signifying potential of the ballet's gestural material.

Mallarmé and Debussy

Nijinsky told journalists that his grasp of the French language was insufficient for him to read Mallarmé's poem (Nijinsky 1912: 418). His ballet, he said, was a response to Debussy's *Prélude*. Debussy for his part, as the title of his piece suggests, intended his music to be a prelude to the poem rather than a programmatic illustration of it. The two volumes of piano *Preludes* that he published in 1910 and 1913 echo this sentiment, as the title of each only appears at the bottom of the page at the end of the music, placed in brackets and following an ellipsis. Gradual suggestion keeps meanings open, as Mallarmé himself was aware: 'The naming of an object suppresses three-quarters of the pleasure of a poem which consists of guessing it bit by bit; suggesting the object, that makes for the dream!' (cited in Butler 1994: 6). The ballet, like the music, was supposed to be a prelude, as a programme note stated: 'A faun snoozes / of the nymphs he tricked / a forgotten scarf satisfies his day dream / the curtain falls for the poem to start in all their memories' (cited Nijinsky 1970: 142). Someone who knew enough about Mallarmé and Debussy must have written this. This may have been Jacques-Emile Blanche or Misia Godebska (Sert), trusted members of Diaghilev's trusted circle who had known Mallarmé well, or Jean Cocteau. However little Nijinsky may have heard about the poem, there are nevertheless correspondences in relation to gender and sexuality between it and the ballet.

What runs through Mallarmé's poem, Debussy's orchestral *Prélude* and Nijinsky's ballet is the use of the imagery from Greek and Roman pastoral idylls to symbolise sexual pleasure. Although the classical theme of Mallarmé's poem resembles those favoured by the conservative Montparnasse group of mid-nineteenth-century French poets, Mallarmé's allusive treatment of sexuality has much more in common with Baudelaire's *Fleurs du Mal* and with the low-life subjects of some of Manet's early paintings – both of which Mallarmé championed in the press. Verlaine included Mallarmé in his list of 'les poèts maudites' (accursed poets) because of the apparently insouciant way in which his poems touched on scandalously erotic material. It is generally accepted that the deliberate but seemingly casual overloading of sensuous language in the poem *L'Après-midi d'un faune* creates, in places, an ambiguity about meaning that pushes to their limits conventional notions of poetic lyricism. Does Mallarmé's faun really encounter nymphs or is it all merely an erotic daydream? And how many nymphs are there? Is there one flirtatious, sensual nymph and another chaste one, or are these two aspects of the same nymph? In the poem's final line the faun addresses them as a couple, but then uses the familiar, singular pronoun 'tu': 'Couple farewell; I go see the shadow which you (singular) became' (Couple, adieu; je vais voir l'ombre que tu devins).

Mallarmé's couple becomes seven nymphs in Nijinsky's ballet, one chief nymph and her six companions. The faun's erotic attraction to them and an atmosphere of sensual reverie are common to both poem and ballet. Although nothing much happens during the ballet (as some contemporary critics complained), for the purposes of analysis it can be divided into six sections. In the first section (bars 1–20 of the orchestral score), the faun, lying alone on a rocky platform a little above the stage, plays his flute and eats grapes. Then in the second (starting at bar 21), some nymphs arrive in varying groups. In the next section (bars 32–51), the chief nymph undresses to bathe, taking off her two veils and her dress, piece by piece, leaving only a thin undergarment. This display arouses the faun who comes down and approaches them, startling one of the nymphs who exits. In the fourth section (bars 52–72) the faun and chief nymph walk closely together from side to side of the stage. She guards herself warily against his predatory intentions. Twice, when the music, marked *expressif*, is at its most intense, they stand still for a couple of bars confronting one another until she seems to wilt away from him (bars 55–6 and 63–4). Finally, he cleanly links one of his bent arms through hers, causing her to drop her dress which she was still holding. She seems to hang on him briefly before he disengages (see Figure 3). In the next section (bars 73–94), the nymphs all leave but the faun, having taken possession of the abandoned dress, repeats on his own the floor pattern of his side-to-side duet with it draped over his elbow

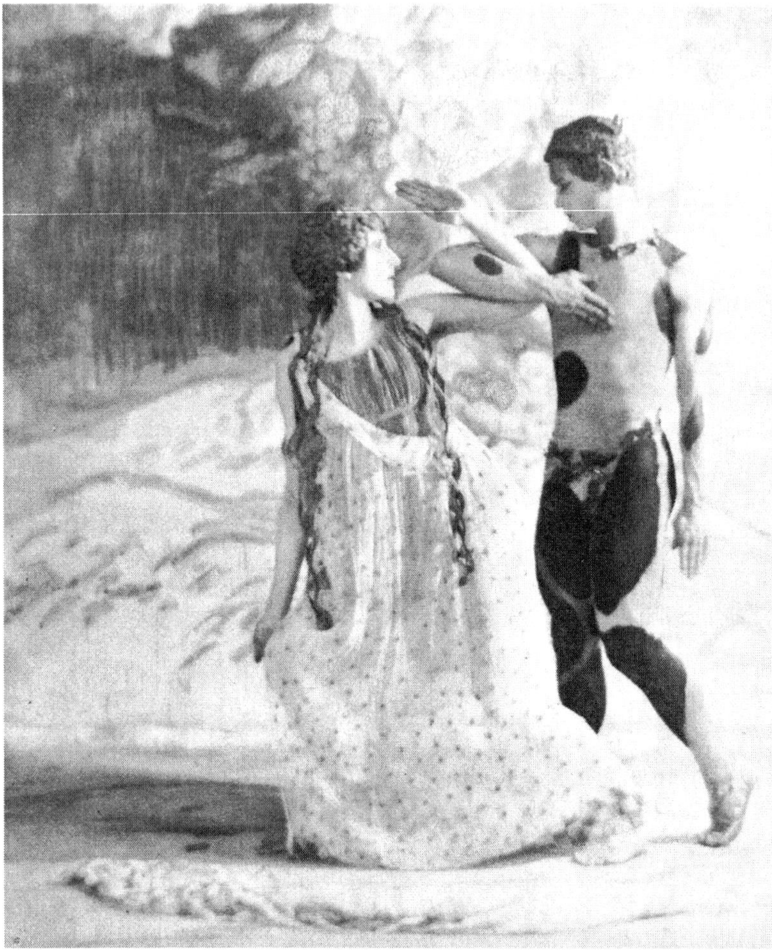

Figure 3 Nijinsky and Tchernicheva in *L'Après-midi d'un faune* (1913)
Photo: Baron Adolf de Meyer. Courtesy of Jerome Robbins Dance Division, The New York Public Library for the Performing Arts, Astor, Lenox and Tilden Foundations.

joints. Two pairs of nymphs return briefly to scold him with sharp little slicing gestures. Finally, in the last section (bars 95–110) he returns to his platform, hugs the veil amorously and rubs it against his face and body before laying it out and lying face down on top of it. In the final few bars of the ballet, the dancer lifts his head, arching his upper back thus pushing his pelvis against the dress, his mouth open in a silent laugh – recalling his earlier laugh after eating his grapes. This might be an orgasmic sigh or he could be laughing triumphantly over his clever capture of the nymph's dress. It is in the final bars of the music, during the first performance on 29 May 1912, that Nijinsky may have put his hand on to his crotch in a way that suggested masturbation.

Debussy's *Prélude à l'Après-midi d'un faune: églogue pour orchestre après Mallarmé* (to give it its full title) evokes the sensuously overcharged but open-ended and ambiguous ambience of Mallarmé's poem. It did so with what were, in 1894, revolutionary uses of orchestral colour. Pierre Boulez described Debussy's reading of Mallarmé as the 'awakening of modern music' (Boulez 1995: 211). Mallarmé had unsuccessfully submitted an early version of *L'Après-midi d'un faune* as a verse play to the Théâtre Français. In the early 1890s, Mallarmé asked Debussy, who was a younger member of his artistic and literary circle, to collaborate with him on a new theatrical presentation of *L'Après-midi d'un faune*. Around this time, Mallarmé was working on the essays on theatre, ballet, and on Loie Fuller that were published in *Crayonné au théâtre*. Debussy initially planned to write three sections, but in the end only wrote the first. As Christopher Butler suggests, the *Prélude,* like Mallarmé's poem, 'sacrifices plot for mood, and lacks the (conventional) harmonic thread of discourse' (1994: 11). Its use of orchestral colour, its fragmentation and regeneration of harmonic and melodic particles, Butler suggests, 'amount to a distinct "Impressionist" style, which is peculiarly sensitive to minute shifts of feeling, and so [is] part of an innovatory reaction against the ossified formal prescriptions of the past' (ibid.). The ballet, too, has a minimal plot in which, some critics suggested, nothing really happens, while its radically stripped-down movement style was clearly also an innovatory reaction against the ossified formality of the classical ballet tradition.

Debussy liked neither Nijinsky's *Faune* nor the ballet *Jeux,* for which he was subsequently commissioned to write the music and which was also choreographed by Nijinsky. Misia Godebska, in her memoirs, wrote that Debussy walked out in disgust from the dress rehearsal of *Faune* (Gold and Fizdale 1979: 143). It was still unusual at the time for dancers to use serious concert music rather than more lightweight 'ballet music' written programmatically for it. Ballet was not something that, up until then, Debussy had had any reason to consider a serious art form. In a letter in June 1913, he expressed to Robert Godet his displeasure at the ballet *Jeux,* complaining at Nijinsky's use, in creating the movement, of the Eurythmics of Jaques-Dalcroze. This was not, of course, true, although Marie Rambert, one of Jaques-Dalcroze's assistants, had been hired to help Nijinsky teach the dancers the complicated, irregular time signatures in both *Jeux* and Stravinsky's *Sacre du printemps* which were first performed at the end of May 1913. Debussy acerbically described Nijinsky as someone who: 'adds up demi-semi-quavers with his feet, checks the result with his arms and then, suddenly struck with paralysis all down one side, glares at the music as it goes past' (Lesure and Nichols 1987: 272). In a somewhat perverse, comic way, this is not altogether an unfair description of the process that dancers and choreographers go through when fitting movement to the complicated and irregular rhythms of innovative twentieth-century music.

Debussy evidently didn't understand or appreciate the kind of musicality that ballet dancers at the time were expected to develop. Lydia Sokolova, who learnt the role of one of the nymphs in 1913, later recalled:

> The dancers had to be musical as well as rhythmical and it was necessary to relax and hear the music as a whole; it had to trickle through your consciousness, and the sensation approached the divine. One walked and moved quite gently in a rhythm that crossed over the beats given by the conductor. At every entrance one made – and there were several – one began to count, taking the count from another dancer who was coming off. For every lift of the hand or head there was a corresponding sound in the score. It was most ingeniously thought out.
>
> (Sokolova 1960: 40)

What Sokolova describes here was a new and innovative relationship between dance and music. Dance steps did not match the music beat for beat as in ballet and much Western folk and social dance up until that time, nor were they danced off the beat as in jazz syncopation. As Sokolova describes it, the phrasing of the dance movement fits into cadences in the music and synchronises with it at key moments so that it is simultaneously both with it and independent of it. As such, it articulated a distinctly modern conception of time (I will return to this and to Eurythmics shortly).

When Sokolova says the sensation of the music approached the divine, she is perhaps referring to its eroticism. In a preface to the original published edition, Debussy wrote that, at the end of his free musical illustration to Mallarmé's poem, the faun 'abandons himself to the delightful sleep, full of visions finally realised [songes enfin réalisée], of full possession amid universal nature' (Debussy 1951: n.p.). What does 'songes enfin réalisée' imply? What was finally realised, and in what way? Musicologist David Code has argued that the allusive eroticism of Debussy's music should be seen as the composer's attempt to create a post-Wagnerian musical idiom, to move beyond the explicitly orgasmic *liebestodt* in the opera *Tristan und Isolde*. Susan McClary suggests that Debussy's evocation of sexuality challenged mainstream conventions. Wagner's *Tristan*, she notes, 'relies on our understanding cadence as sexual consummation [...] and derives its overwhelming power by delaying for hours its long-awaited closure' (1995: 100). Modern music at the end of the nineteenth century, she argues, 'depends on the questioning of then-conventional images of the body or sexuality: Debussy's faun opted for sustained pleasure rather than climax' (ibid.: 101).[2] Debussy, working in the medium of music, used repetition, variation, fragmentation and circularity in order to avoid conventional closure and thus maintain a heightened atmosphere of sensuality.

Nijinsky, through arranging concrete dancing bodies on stage, could not avoid making explicit certain details about faun and nymphs that had been left ambiguous in poem and *Prélude*. As the faun, he wore one of Bakst's most reproduced costumes: tights and a sleeveless leotard decorated with irregularly shaped round patches of brown (that alluded to goat fur), with a similar patch painted on his bare upper arm and a smaller one on his forearm. This revealing, skin-tight costume would have carried an erotic charge. Huntly Carter, writing about Bakst's designs in *The New Age*, mentions their 'intense feeling for sex' (1912: 211). While this kind of costume has since become unremarkable, it was shockingly daring in its day. What, I suggest, the ballet retained from Debussy and Mallarmé's works was a generalised sense of indolent reverie and unproductive arousal. However, what distinguishes Nijinsky's ballet from the two earlier works was the disturbing, modernist sensibility that is evident in the sometimes negative responses of some contemporary critics to its performance.

The critical reception of *Faune*'s modernism

The first public performance of the ballet *Faune* at the Théâtre du Châtelet on 29 May 1912 triggered off a heated debate in the French press concerning its final moments. Nijinsky seems in some way to have performed a gesture that could have been interpreted as masturbation. Gaston Calmette, the editor of *Le Figaro* suppressed what was expected to be a positive review by the paper's music critic Robert Brussel and printed in its place a column he himself had written titled 'Faux pas', a pun suggesting both a false (ballet) step and a social indiscretion.

He denounced what he saw as Nijinsky's bestial, erotic movements and strongly indecent gestures.[3] Reports of the controversy appeared internationally. It is unclear what Nijinsky may have done during the first performance. According to Romola Nijinsky, after the police became involved, Nijinsky changed the ending (Nijinsky 1970: 145). Diaghilev, an early master of media manipulation, provided the newspapers with letters in support of the ballet. The sculptor Auguste Rodin signed a letter written for him by Roger Marx which was printed in *Le Matin* on 31 May, and one by the painter Odilon Redon was printed in *Le Figaro*. The latter had been a friend of Mallarmé's and had illustrated his revolutionary, late poem 'Un coup de dés'. It is assumed that Diaghilev paid Redon and bribed Rodin by offering Nijinsky's services as model for a sculpture. Redon's letter, nevertheless, confirms the ballet's relationship with Mallarmé's ideas, recalling how Mallarmé was always bringing dance and music into his conversations. His spirit, Redon concluded, had been with them all in the Théâtre du Châtelet that night.[4]

The result of this controversy is that Nijinsky's first ballet received a great deal of press attention in France. Maurice Touchard, who largely agreed with Calmette, suggested that the unfortunate behaviour of Slavic artistry in the presence of Latin genius proved the alliance between Nijinsky and Debussy was ill-matched (Touchard 1912: 206). The general consensus, however, rejected the accusation of indecency, although many felt Nijinsky's ballet did not catch the spirit of Mallarmé's poem or that the choreographer shouldn't have tried to use Debussy's music. There is perhaps an element of national pride involved here, an unwillingness to accept the possibility of a Russian being able to engage on an artistic level with advanced French culture.

After the first performance of *Faune,* Jacques Rivière welcomed Nijinsky's *Faune* and reported that it had stimulated passionate discussions among artists and critics because of its cutting-edge modernism (*modernisme suraigu*) and its completely unprecedented conception (1912: n.p.). Prior to making *Faune,* Nijinsky was celebrated for his spectacular leaps and *coup de théâtre*. Some critics complained that *Faune* consisted mostly of walking and that Nijinsky only leapt once during it. Nijinsky's training within, and mastery of, older artistic styles is a recognised characteristic among early modernist artists like Picasso, Matisse and Schoenberg. As Christopher Butler notes: 'An awareness of the past and the technical ability to reproduce it ensured that the early advances of the modernists were made through a stylistic metamorphosis of the genres of the previous tradition' (Butler 1994: 73). Marie Rambert corroborates this when she states that, for *Faune,* Nijinsky required 'a perfect ballet technique and then broke it down consciously for his own purposes' (1972: 61). Whereas in the Ballets Russes and in the Imperial Russian Ballet at the time talented dancers were encouraged to exercise a degree of freedom in creating personal interpretations of their roles, the 1912 production of *L'Après-midi d'un faune* was the first work in which the dancers all had to conform uniformly to the choreographer's prescribed manner of interpretation.[5]

The floor patterns of the choreography were straight, parallel lines in a narrow track right at the front of the stage; the back cloth for the ballet in the Théâtre du Châtelet in May 1912 was placed only two meters behind the proscenium arch, and the dancers moved on a shallow, painted floor cloth (Guest and Jeschke 1991: 15–16). The dancers' feet are in *parallel* and not *turned out* as in classical ballet. Faun and nymphs walk in a severely horizontal, frieze-like flattened way. Thus, while their feet and legs walked from side to side of the stage, their shoulders were held as near as possibly square to the audience but their face in profile, the head sometimes tilted (see Figure 3). When a dancer came to the end of their path, they swivelled with their feet neatly on the spot and proceeded back the way they had come. Bronislava Nijinska, who performed the role of the joyful nymph in the 1912 production, gives a revealing description of the jump she made when she suddenly encountered the faun:

> The difficulty of this jump was that I had to preserve in the air the same pose as when running across the stage – a bas-relief form, with the knees slightly bent. I had to jump high without any additional preparation, and then on landing I had to make an abrupt half-turn without lifting my heels from the ground, before running away to the wings still maintaining the same bas-relief form.
>
> (1981: 404–5)

Nevertheless, the dancers have to remain as relaxed as possible. As Sokolova recalled: 'In order to preserve the patterns of the frieze, you had to keep your hands and arms flat in profile: to do this it was necessary to relax the hand and arm, for if you forced or tightened the gesture, the wrist fell back and the straight line from elbow to fingers was lost' (Sokolova 1960: 41). Claudia Jeschke sums up *Faune* innovations as follows:

> The groundedness of the dancers' movements, the angular rather than rounded postural imagery, the absence of balletic virtuosity,

> the shift of action from the legs, so essential to classical ballet, to the upper body and arms, the denial of visual synonymy between musical pulse and movement, the avoidance of melodrama, and the refusal to signal, through mime or pantomime in such a way as to predetermine the audience's reaction.
>
> (Guest and Jeschke 1991: 116)

To this should be added its compression of stage space, its dancers' refusal to meet the audience's gaze, and its stylised reduction of conventional balletic virtuosity.

This modernist way of moving attracted criticism. Adolphe Julien, music critic of the heavy-weight *Journal des Débats*, responded dismissively to the proposal that *Faune* was a cubist dance: 'This attempt to bring to life antique friezes with jerky, almost contorted gestures, and abrupt, automatic movements with unexpected pauses, has nothing to do with living dance at all but looks as if it is being performed by mechanical puppets' (Julien 1912: 2). In a similar vein, Emile Vuillermoz regrets the detestable progress of intellectualism that has caused Nijinsky, in trying to recreate Greek vases, to study the work of Metzinger and Picasso and 'isolated the human rhythm of the bodies displayed in the Salon des Indépendants' (Vuillermoz 1912: 65). Despite Diaghilev's suggestion, there is surely nothing stylistically cubist about *Faune*. These two writers were citing recent controversies about Fauve and Cubist painting exhibited in the annual Salon des Indépendants in order to criticise *Faune*. Julien was disturbed by what he saw as the ballet's fragmented, machine-like quality. Vuillermoz, whose conservatism dictated that ballet could not be intellectual, could not countenance the demand to actively engage with, and make his own sense of, a challenging, innovate work. As I noted earlier, what was unsettling about artistic modernism was the loss of the familiar and its replacement with absences that reminded beholders of their own experience of modern living.

Faune's modernism was well received by other writers. Louis Schneider wrote that it opened up 'a completely unexploited region in the art of choreography' (1912: 9). Schneider suggested that the style of movement Nijinsky had developed was analogous to that of Jaques-Dalcroze. Diaghilev and Nijinsky, in fact, visited the latter's School of Eurythmics in Hellerau for the first time a few weeks later. As Schneider recognised, other modernist movement styles were being developed around this time. One such was emerging in the work of the theatre director Vsevolod Meyerhold. In St Petersburg in 1910, Nijinsky and Meyerhold had both created roles in Fokine's *Carnaval*. Lynn Garafola suggests that Nijinsky was influenced by Meyerhold, who had developed a static, motionless style of performing Symbolist plays around 1905–8 in which actors moved in flattened, frieze-like groups. She points out that Bakst had collaborated with Meyerhold on several theatre projects and that Diaghilev also knew the director (Garafola 1989: 30–2, 53–5). Although it is unlikely that Nijinsky would have seen any of Meyerhold's productions and hadn't yet seen any Eurythmics, these examples show that what Nijinsky discovered through making *Faune* corresponded to other modernist forms of movement research being carried out by other dance and theatre innovators during this period.

Underlying this innovative approach to the human body in motion is a new way of conceptualising the experience of time to which I referred earlier. I noted that the counts of the dance sequences in Nijinsky's choreography occur alongside, but relatively independent from the complex, continually changing time signatures in Debussy's music. In 1919, Nijinska, marooned by the civil war in Kiev and cut off from her brother, was writing about his dancing in a treatise on dance pedagogy. In her judgment, whereas conventional ballet schools only taught separate movements, Nijinsky and Pavlova had discovered the secret of thinking and acting between positions:

> Do you remember how Pavlova approached an arabesque? This is how one should learn movement. Or Nijinsky? Do you remember how many transitions there were in the course of his leap? These transitions and nuances created the illusion that he never touched the ground.
>
> (Nijinska 1986: 86)

It is a sensitivity to such transitions and nuances during the evolving moment of performing the ballet that, I suggest, the dancers in *Faune* exemplified.

The difference between dancing separate steps in a regular, rhythmical way and dancing across and between the beats resonates with Henri Bergson's theorisation of a multiplicity of simultaneous temporalities. In Bergson's terms, the dancers and the musicians in *Faune* were each counting in different rhythms. In *Matter and Memory* Bergson proposed that, in Greek philosophy, a dialectical succession of privileged instants was understood to actualise transcendent forms. In modern conception, time is not divided up in any predetermined way: 'We may divide it [...] in any way we please' (1988: 192). Bergson contrasts the fixed way in which movement is conventionally presented in the Parthenon Frieze with Muybridge's series of time-lapsed photographs of a galloping horse. Whereas the former represents one ideal, privileged moment, in the latter, he suggests, 'all moments count equally' (ibid., 'tout les instants se valent') and none is any more important than any other. All moments count equally in *Faune*.

Matter and Memory drew on the latest ideas from experimental science and cognitive psychology to demonstrate the shortcomings of a rational, scientific, deterministic way of conceptualising time and memory. Ultimately it argues that modern science cannot capture the rhythm of modern life. Whether or not *Faune* captured the rhythm of modern life was, as I have shown, an issue that some critics discussed. It is hard for a twenty-first-century viewer to understand how Julien and Vuillermoz could find the movement in *Faune* jerky. They perhaps expected a 'musical visualisation' of the *Prélude*'s textures and flow. Instead, they were presented with a modernist fragmentation of rhythm, the absence of any regular set of clearly identified, fixed moments. This was the result of Nijinsky's stripping back of the conventions and techniques of classical ballet to a bare formal language. In Raymond Williams' terms, this modernist reduction relates to the experience of modernity exemplified by the life of the modern metropolis. Where some compared *Faune*'s movement style with Jaques-Dalcroze's exercises, they were suggesting that it defined a human rhythm appropriate to modern times. Paradoxically, just as Bergson used modern science to demonstrate the limitations of a scientific approach, *Faune* broke with the past, and with past signifying practices, in a way that evoked modernity, in order to capture human movement qualities which many feared modernisation was in danger of eroding.

This break with the past is particularly evident in the ballet's lack of action and the absence in it of any conventional signs that might suggest a narrative. As in Muybridge's photographs, no one moment in *Faune* is any more significant than any other. No instant can therefore stand out as having any particular meaning because the meaning is spread out throughout the ballet as a whole, just as the gallop spreads out across the whole series of photographs. The absence of privileged, conventionally meaningful moments in *Faune* means the disappearance of any opportunity in which to display a significant gesture. While creating the ballet, Nijinsky became a collector of gestures which he sought in archaic Greek vases in the Louvre and the British Museum, but did not use them in the way mimetic signs were used in nineteenth-century ballets to create narrative meanings. In

Faune, meanings are conveyed through the expressiveness of the body as a whole. Thus, for example, when the nymphs scold the faun, it is the energy and intensity of their slicing movements that convey their intention rather than any coded meaning signified by the particular shape their hands and arms take up. Throughout, the choreographed inclinations of the head convey more narrative information than any gestures made with hand or fingers. The Italian philosopher Giorgio Agamben proposes that, by the end of the nineteenth century, the European bourgeoisie had lost its gestures (2000: 49). Citing the dances of Isadora Duncan, the work of the Ballets Russes and the early silent cinema as examples, Agamben argues that the particular fascination such work had at that time exemplified humanity 'trying for the very last time to evoke what was slipping through its fingers' (ibid.: 54). There is a significant difference between Fokine's choreography for the Ballet Russes, which had made Nijinsky famous, and the innovations the latter introduced in *Faune*. Fokine tried, in Agamben's terms, to evoke the memory of gestures that were already slipping away, while their absence in *Faune* emphasises the impassable barrier between past and present that, in Walter Benjamin's view, modernity had created. This absence was affective.

For this reason, I think it is unlikely that Nijinsky actually made any masturbatory gesture at the end of the ballet, although his performance undoubtedly suggested extreme sensuality and eroticism. Hanna Järvinen (2003) has drawn an extremely useful distinction between 1912 French and Russian attitudes towards masturbation. For Western Europeans it would have been unthinkable to talk openly about masturbation; Russian critics, writing from Paris, were puzzled by the scandal surrounding *Faune*'s first performance. Järvinen quotes Nikolay Minsky, who wrote:

> No doubt the image is daring, even opposite to [our] understanding of the fauns of Antiquity, but no doubt it is pornographic only to the lustful imagination, in contrast to Nijinsky's naive, honest depiction of something that 99 out of 100 young men has got himself acquainted with.
>
> (cited in Järvinen 2003: 215)

Järvinen points out that while Russian beholders may have found *Faune* shocking in an aesthetic way, they seem not to have found it perverse or morally degenerate in the way that some of their French counterparts clearly considered it to be.

In Western Europe and North America, masturbation was an unmentionable subject because, as Michel Foucault has pointed out, sexuality had, since the seventeenth century, been systematically turned into a field of discourse. Masturbation, particularly among young adolescents, was something that Foucault suggests had been forced into hiding in order to make possible its discovery (1979: 42). Thomas Laqueur proposes that, underlying nineteenth-century treatises on the dangers of onanism was a fear of unproductive idleness. Surprisingly, he goes on to connect this with medical concerns in the eighteenth century of the effects of excessive reading of novels. Discounting the moral and pathological arguments for abstaining from masturbation, Laqueur proposes an underlying fear of the imagination and of 'a satisfaction of endless desire by endless gratification' (1995: 126). I have shown that a mood of erotic reverie or daydream pervades the poem, the *Prélude* and the ballet. As Jonathan Crary has argued, a daydream is 'a domain of resistance internal to any system of routinization or coercion' (1999: 77).

If Gaston Calmette was partially successful in using *Le Figaro* to turn Nijinsky's masturbating faun into a phobic creature, this was because the ballet evoked a mood of erotic reverie that was potentially uncontrollable and thus threatened the values of productive rationality on which the modern state depends. Compared with contemporary Western European attitudes towards sexuality, Russian anxieties about sexuality had not, in Foucault's terms, become subject to institutional investment. This different conceptualisation of sexuality left Nijinsky free to bring out associations implicit in Mallarmé's writings, particularly in relation to the formation of gendered subjectivity.

Absence and gendered subjectivity

In Mallarmé's two essays on dance in *Crayonné au théâtre* – on La Cornalba in the ballet *The Two Pigeons*, and on Loie Fuller – the female dancer is either figured as someone who creates something out of nothing or as disappearing in the act of creating signs. The female dancer is thus very like the *khora* in Plato's *Timaeus*, an empty, generative space. In his essay 'Ballets', Mallarmé said that the dancer is not a woman who dances nor even a woman, but a 'metaphor' for the ballet's imagery. She does not dance but 'suggests, by way of prodigious abbreviations or accelerations (suggérant, par le prodige de raccourcis ou d'élans), with a corporeal writing that would necessitate paragraphs of dramatic dialogue as well as prosaic description, to be expressed, in the rewriting: poem disengaged from all of the scribe's devices (poème dégagé de tout appareil du scribe)' (Mallarmé 1976: 192–3). Running through this difficult passage is a teasing tension between the presence of dancing and the absence of the female dancer, between corporeal materiality and aesthetic signs that, through their immateriality, signify more than written words can convey. By ceasing to be a dancer and thus becoming nothing, the dancer writes a corporeal poetry. This becomes part of the exchange between performer and beholder when, later in the same essay, Mallarmé described the ballerina as an unconscious revealer of truths who: 'through a commerce whose secret appears to spill from her smile, without delay, she delivers to you, across that last veil which always remains, the nudity of your concepts and silently writes your vision, like a Sign, which she is' (ibid.: 197).[6]

Mallarmé restated his fascination with the play, in dance performance, between presence and absence in a famously untranslatable essay about Loie Fuller: 'The enchantress creates the ambience, draws it out from herself and returns into it (la tire de soi et l'y rentre), in a palpitating silence of crêpe de chine' (ibid.: 200). The chief nymph in the ballet *L'Après-midi d'un faune*, by divesting herself of her veils, might similarly be said to create a feminine space for herself in a palpitating bath of silk. Through playing with the similarity between the French word 'soi' (herself) and 'soie' (silk), Mallarmé evokes an image of the dancing space of silk which Fuller creates around herself as an expression of her feminine interiority. Felicia McCarren links Mallarmé's poetic allusion here with contemporary medical discussions of hysteria. At the time, it was believed that (female) hysterics manifested physical symptoms of illnesses by imitating physical causes from which they were not actually suffering. The word hysteria comes from the Greek word *hystera* meaning the womb. Like the womb, the hysteric was thought to create something out of this emptiness. As McCarren observes: 'The subject of Fuller's dance, then, for Mallarmé, is the female body, the inside space of that body, the theatre within – the creative space of *hystera* the womb' (1995: 758).

Superficially Mallarmé's description of a female dancer who becomes an absent sign through the act of performing conforms to nineteenth-century gender ideologies. In these, woman was associated with the lowly realm of the body and nature, while man was associated with a higher sphere comprising disembodied culture and metaphysical reason. Penny Florence, however, proposes a more complex reading of Mallarmé's later poems that is also applicable to his essays on dance:

> In Mallarmé's later poems women merge with the poetic process, partly through transference of his codes for poetry (whiteness, flight,

> light, potential utterance etc.) to women, and women's bodies (e.g., in the triptych 'Tout orgueil…', 'Surgi de la croupe…', 'Une dentelle…'). This transference exemplifies the slippage that often occurs between the woman and other meanings. It is a process that may be read as eradicating the woman, and this is often the case. But it may also open up the 'female' as a complex site of meaning which may be read as centring women.
>
> (1986: 129)

It is my contention that the faun's flat, frieze-like movement in the ballet *L'Après-midi d'un faune* opens up the male dancing body as a complex site of meaning. By doing so, it creates a comparable slippage between masculinity as an unexceptional metaphysical ideal and a shocking assertion of the irrational allure of a desiring and desirable masculine body. This slippage opens up normative, unproblematic ideas about masculinity into a more complex and problematic site of meaning.

Whereas perspective locates the figures in paintings within space, allowing them to be surveyed from a dominant viewpoint, the ballet's flattened, frieze-like lack of perspective in effect denied the beholder the possibility of fixing the dancers in this privileged manner. As Claudia Jeschke notes:

> By moving his dancers almost exclusively from one side of the stage to the other, he evolved a choreographic structure that allowed not one but several centers of spatial and theatrical gravity. As a result, he avoided the illusionism of centralised perspective, an illusionism that treats space as if it were real space in which quasi-real events unfold.
>
> (Guest and Jeschke 1991: 104)

What links Nijinsky's performance of sexual behaviour with the performance of his frieze-like, flattened movement material is the way that both trouble the relationship between performer and beholder in a disturbingly deconstructive way. Like Loie Fuller in Mallarmé's reading, both the male and female dancers created themselves out of nothing. Following Florence, for a woman to create herself out of nothing was to problematise the philosophical tradition which confined women to the physical realm and denied them access to a higher metaphysical one. For a male dancer to produce himself out of nothing was to destabilise the grounds on which male power depends. It troubled the idea that masculinity is unproblematic. Even more disturbing, this self-production was not an act of masculine striving, but a seemingly passive by-product of an indolent, sensual, erotic daydream. Masculinity only remains a disembodied norm as long as men remain invisible and unmarked. By drawing attention to his own arousal, the faun forfeited the status of normatively unproblematic masculinity. Because his role would not remain invisible, the ballet enacted a subversive rupture within normative ideologies.

In conclusion, while Nijinsky's experience of the strangeness of the foreign metropolis may have contributed to *Faune*'s radical examination of the medium of dance movement, it was Diaghilev's organisation that ensured the ballet was discussed in the terms of the exciting new critical vocabularies of modernism. Modernism, as I have argued, was ambiguous and potentially contradictory in the way it attempted to critique what were in effect the conditions of its own possibility. This ambiguity opened up possibilities for troubling and subverting nineteenth-century gender ideologies. The ballet's acceptance by European cultural elites in effect legitimated its daring eroticism. Nijinsky's Russian approach to sexuality led him to draw attention to corporeal aspects of masculinity that problematised the nineteenth-century Western European association of masculinity with disembodied reason and culture. It made explicit what had been implicit within the male balletomanes' erotic fascination with female dancers (which Valerie discussed in Chapter 1). Like Mallarmé's late poems and his essays on dance, the ballet broke with notions of rational unitary, universal subjectivity, suggesting instead the historically specific and non-cogent, non-unified processes of masculine and feminine subject formation. The faun's role in the ballet thus revealed these disruptive and irrational aspects of contemporary masculine subjectivity through the use of imagery and aesthetic sensibilities that were developed by Mallarmé and Debussy; but it did so by making explicit in a troublingly deconstructive, modernist way what had been implicit but kept veiled by these older artists.

Notes

1. An earlier version of this chapter appeared in Italian as 'Maschilità decadenti e L'Après-midi d'un faune', in Marco Pustianaz and Luisa Villa (eds) *Maschilità decadenti: La Lunga fin de siècle*, Bergamo: Bergamo University Press (trans. Patrizia Verolli), pp. 337–55. I am grateful to Marco Pustianaz for his comments at the time. I am also very grateful to Hanna Järvinen not only for her generous feedback when I was rewriting the chapter for this book and giving me a copy of her PhD but also for sharing with me her copies of contemporary French reviews of Nijinsky's production of the ballet.

2. See also my co-collaborator Valerie Briginshaw's discussion of McClary on climax in classical music in Chapter 1.
3. 'mouvements de bestialité erotiques et des gestes de lourd impudeur', *Le Figaro*, 30 May 1912.
4. 'L'esprit de Mallarmé etait ce soir parmi nous', *Le Figaro*, 31 May 1912.
5. Bronislava Nijinska pointed out that: 'Up to then the ballet artist had been free to project his own individuality as he felt [...] Nijinsky was the first to demand that his whole choreographic material should be executed not only as he saw it but also according to his artistic interpretation' (Nijinska 1981: 427).
6. The teasing tension here is very different from the one Valerie has identified in Gautier's writings (in Chapter 1); indeed Mallarmé's comments about the nudity of one's concepts might even be read as a veiled criticism of such fetishistic writing.

Section 2

Rethinking Temporality: Preposterous Histories

Introduction

This section includes three chapters, one that arose out of a writing and performance-making research project entitled *Embodying Ambiguities* with the choreographer Emilyn Claid, and two others which each compare a canonical ballet with a piece of experimental dance that belongs to different historical periods. The reason for grouping these three chapters within this section is that they are each concerned with rethinking temporality. The first reconceptualises time as duration, drawing on theories of Deleuze and Bergson. The other two disrupt notions of linear progressive time that underpin normative historiography, in order to create new anti-canonical ways of conceiving dance history together with new insights into particular works. This is one instance where our different aims and methods have resulted in chapters that have prompted a common realisation of the need to rethink temporality. Whereas Valerie and Ruth Chandler's chapter explores notions of time conceptually through writing and experiencing one dance work, Ramsay's chapters both make what he terms a preposterous comparison, looking at history the wrong way round. Valerie and Ruth Chandler discuss movement that disrupts logical linear ways of thinking and seeing performative events in order to open up and discover new ways of re-linking these on a different, non-hierarchical level. Ramsay's approach is to re-link ideas and experiences of travel, humour or popular social dances that were the sources and inspirations of the dance works he discusses.

3

Rethinking Temporality: Intertextual Plays Within and Between Discourses of Space, Time and Performing Bodies

Valerie A. Briginshaw and Ruth Chandler

Introduction

The original version of this chapter arose out of the *Embodying Ambiguities* Arts and Humanities Research Board funded writing and performance making project (1999–2004), co-directed by Valerie and the choreographer, Emilyn Claid. This involved exploring the ambiguities between different concepts of space and time at play in live dance performance. It concerned an interweaving of creative interactions between five areas of activity: the thinking of choreography; the embodied physicality of performance; observation of the resulting movement material; reading related philosophical texts concerned with reconfiguring spatial and temporal discourses; and writing texts that derive from and further articulate these experiences. The project produced various written papers and a trilogy of dance works.[1] The first, *Shiver Rococo* (1999), because its various layers of choreographical exploration inhabited and played between seemingly dissonant perspectives, stimulated and provoked us to use the work of Deleuze and Bergson concerning the ways in which traditional notions of space and time are reconfigured. The focus of this chapter, which is a reworked version of a paper Valerie wrote with Ruth Chandler, another writer on the project,[2] concerns rethinking temporality in particular[3] and the relations with space that involves.

'Rethinking' and 'temporality' are the key words here because we are interested in exploring the ways in which the work of Deleuze and

Bergson can make us think dance and experiences of temporality differently.[4] We are not concerned with trying to encapsulate the dance in writing. We do not believe this is possible. Rather, we are concerned with showing how dance can help in creating thought, in creating new ways of thinking, specifically new ways of configuring time. The philosophies of Deleuze and Bergson are concerned with multiplicities, with pluralisms, with many things occurring simultaneously, recognising that human beings and the world consist of multiples. These philosophers also see thought or intelligence – or ways of thinking or interpreting things – as multiple, involving, as they do, perceptions, affects, intensities and feelings or sensations, which are also inevitably pluralistic or multiplicitous. Seeing things as simultaneously multiple involves the recognition that they are continually in flux or moving. The continuous flow of movement is key. Understanding movement in this way foregrounds its immanent dynamism, force or energy; its power or potential for action. It also involves understanding movement not as happening at instants in time but in the intervals in between, where perceptions, affects, intensities and sensations fluctuate, hence the potential. Seeing things in this way means understanding time not as linear, chronological and measurable, but, as we experience or sense it, as duration. Dance, consisting as it does of movement, allows us to explore the potential of these in between intervals, that allow us to rethink temporality in this way. Our encounters with the movement of *Shiver Rococo* suggested exploring this in between potential by way of three interrelated notions from the theories of Deleuze and Bergson: 'irrational cuts', objects becoming 'objectile', and 'affective openness'. Multiplicity and pluralism interweave through these three, creating images of thought – or thinking – that move and are dynamic, and images of time as duration; a non-homogenised time, a time 'out of joint' that disturbs habit and has 'a potential for difference' (Colebrook 2006: 27). These images of time also inspire Ramsay's two chapters that follow in this section since they inform ways of rethinking history.

The initial research for the project consisted of daily movement research sessions over four weeks in 1999 for the choreographer, Emilyn Claid, and five dancers (Yalckun Abdurehim, Matthew Hawkins, Stine Nielson, Sue Smith and Lisa Turon), and reading and writing conducted during and between five of the movement research sessions observed by the writers. Claid had chosen her dancers for their highly developed technical skills and their dance backgrounds. Matthew Hawkins, for example, was trained at the Royal Ballet but had also worked in various experimental dance contexts. Claid had worked extensively with Sue Smith, who danced with *CandoCo*, and admired not only her technical facility but also her intelligence as a dancer and her ability to improvise. Claid was also attracted to Yalckun Abdurehim's strong technique and his distinctively sensitive almost androgynous way of moving. Stine Neilson and Lisa Turon both possessed a strong sense of line and bodily presence, and in Turon's case this was arrestingly combined with quite a slight build. The five dancers were all very different and individual in appearance, in their training, and in their distinctive movement qualities and reactions. What they shared was virtuosity and a considerable facility for moving in a range of dance styles underpinned by highly developed technical abilities.

Claid relished the luxury the project gave her to conduct movement research over a period of time without having the pressure of having to ultimately produce a finished dance piece. This resulted in around 20 minutes of movement material accompanied by a devised musical score by Stuart Jones, also still in its formative stages of composition, which was shared publicly, without any special lighting or costume, on three occasions with two university audiences, and with a group of other artists and friends. Not having to produce a finished piece of choreography meant that the resulting movement material remained refreshingly open and the results of Claid's main research aim, which was to explore movement that played in ambiguous spaces between extremes of dance languages, were immediately visible. Having chosen to research movement between dance languages, the result was abstract and plotless; Claid was not concerned with narrative or character. The movement material for *Shiver Rococo* played between movement styles as diverse as ballet and butoh, where technique and improvisation were equally valued, and where a strong sense of line coexisted with distorted and fragmented linear designs. It often resembled work by choreographers such as Cunningham or Forsythe. In the dance arena, bodies are normally seen to alternate spatially between prominent and peripheral activities, but Claid's choreographic research for *Shiver Rococo* was structured to be observed from all angles with the audience seated in a rectangle surrounding the performance space. It decentred the traditional centre–marginal geometry of the traditional Western theatre dance 'scene'. This was one example of the ways in which this dance research disrupted conventional thinking in terms of binaries and instead enabled seeing, perceiving and experiencing the movement from shifting perspectives.

Irrational cuts

In his book, *Bergsonism,* Deleuze insists that, to avoid false errors of differences in degree and kind between space and time, problems must be stated in terms of time rather than space (1991: 31). This is because time or duration bears 'differences in kind' whereas space only presents 'differences of degree' (ibid.). The being of time is a multiplicity of lived durations of which only a few can become discernible to humans. Bergson's total virtual past posits a temporal monism made up of 'an infinity of actual fluxes' (ibid.: 82) or a generalised pluralism. All fluxes participate in the same indivisible virtual whole, but their individual action is that of a limited pluralism, 'dualism is therefore only a moment, which must lead to the re-formation of a monism' (ibid.: 29). Deleuze's last return to Bergson – his studies of cinema – suggests that twentieth-century film history documents an aesthetic shift from what he calls movement/action images to time images. He argues that from a notion of time as pre-existent to their image possibilities, filmic techniques have moved to produce time images – virtual/actual crystals – which collapse any fixed notion of time or space as categorically *a priori* to their potentials. Deleuze moved from movement images to time images. Our rethinking of dance here moves us from time images to rethink movement images. With Deleuze, we challenge the classic image which supposes that 'movement in its extension was the immediate given' (Deleuze 1989: 277) and which we see in much Western theatre dance. Instead of the 'rational cut', where images are 'linked or extended according to laws of association, of continuity, resemblance, contrast, or opposition' (ibid.: 276), Deleuze suggests the irrational cut as 'noosign: an image which goes beyond itself towards something which can only be thought' (ibid.: 334). The irrational cut is not a second moment which follows a first, like a line, but an incommensurable 'limit or interstice' (ibid.: 278) between non-linked but re-linked images. The images are non-linked in that they bear no logical relation to one another and re-linked through the technologies of film and of our interpretations of them, putting them together. 'From this confrontation "thought" appears' and thought that is not closed, but pluralistically open (ibid.: xvi–xvii). Analysing *Shiver Rococo* in terms of irrational cuts demonstrates the complexity of the choreographic and performance plays at work and reveals some of the ways in which they prompt pluralistically open thought and, in the process, reconfigure notions of time.

Drawing on Bergson's thesis that there are two forms of recognition, Deleuze suggests that 'the present is the actual image, and *its* contemporaneous past is the virtual image' (1989: 79, his emphasis). These two aspects to crystal images form a 'whole' which is in continual variation and 'memories, dreams, even worlds are only apparent relative circuits which depend on the variations of this Whole. They are degrees or modes of actualisation which are spread out between these two extremes of the actual and the virtual' (ibid.: 81). Cinema, Deleuze suggests, 'is a new practice of images and signs, whose theory philosophy must produce as conceptual practice. For no technical determination, whether applied (psychoanalysis, linguistics) or reflexive, is sufficient to constitute the concepts of cinema itself' (ibid.: 280). Something similar may be said of relations between danced images in some radical innovative dances such as *Shiver Rococo.*

Despite Deleuze's analysis coming from his study of modern cinema we are arguing that it can inform our thinking about some dancing, since he suggests that the 'irrational cut may present itself in quite diverse visual forms' (1989: 248). We are not suggesting a simple transposition of the time image on to movement images of *Shiver Rococo*. Nor are we proposing to collapse the distinction between extended and intensive repetition which supports the actual and virtual. Rather, our contention is that the irrational cut of film can be embodied in dance for images that do not seek a correspondence between their plane of expression and the movement of time. In so doing, it is possible to dialogue with movement images that express the habitual story forms of representational logic and to undo their normative narrative effects and repeat them differently. By doing this, it is possible to rethink history as multiple planes and mediums of expression.

Deleuze suggests that the sound image in cinema 'frames a mass or a continuity from which the pure speech act is to be extracted' (1989: 279). It is here that the crystal is born as a direct time image in the perpetual relinkages of irrational cuts. The visual and sound images exist in 'a new asynchrony' (ibid.: 250) and we are suggesting this is the case with some danced images. They no longer correspond or are 'held together', they contradict themselves and these contradictions 'induce a system of unhookings and intertwinings which in turn determine the different presents through anticipation or regression, in a direct time-image' (ibid.). According to Deleuze, 'the visual image for its part frames an "any-space-whatever", an empty or disconnected space which takes on a new value' (ibid.: 279). Drawing on the techniques specific to film, Deleuze suggests that 'the visual image will thus never show what the sound image utters', but it is the 'irrational cut' passing between visual and sound images which makes the consistency of film. Similarly, a danced image can never be translated into words and described. The relations between the formation of the danced images and our visual and written images of *Shiver Rococo,* and between some danced images within the piece, can be seen to work in a manner similar to the 'irrational cut', in that many of the danced images in their shifting ambiguities and incommensurabilities seem to emerge at the border of thought or at the edge of reason. Freezing thought at the edge of reason into the consistency of stills plays on the one hand with representational logic. Movement is immobilised, pinned down and framed as an object. On the other hand, the stills may be regarded as snapshots of transition[s][5] of movement that are never still and always elsewhere. There is no point at which objects may be fixed in a place. What we see in our reworked movement images are images that achieve perceptual consistency as shapes but which are not framed by the dream of an absolute horizon in the way representational images are.

One such example of a danced image of this kind resonates with Deleuze's evocative description of hallucinations: 'That we were perceiving in folds means that we have been grasping figures through the haze of dust' (1990: 4). There is a pivotal moment of unison movement in *Shiver Rococo* when all five dancers are sitting on the floor in the same position, legs stretched out in front of them, their linear formation dividing the performance space. Although the dancers are still, they are not stopped. They are rather in a state of what Erin Manning terms 'preacceleration' or 'incipient action', 'felt as the virtual momentum of a movement's taking form before we actually move' (Manning 2008). Matthew, in the centre, breaks out of the group, swaying his weight to one hip and then the other, and swinging his body over to face the ground, balancing on his feet and one hand, leaving his other arm swinging freely. This swinging seems the only movement in the space. It appears magnified, yet impotent. Despite its size and momentum, it has no effect on the other dancers, who continue to sit looking in the opposite direction apparently resolutely ignoring Matthew's 'break away' gestures. Magnified, yet impotent, his gestures create a gap or interstice in which the relations between himself and the incipient action of the others appear incommensurable, disjunctive.

Matthew then gets up and 'purposefully' walks away from the line to the audience on the periphery. Standing facing them, tall and proud, he stamps his foot forcefully. Spreading his arms wide, he jumps slapping his hips with his hands. Whilst landing he makes an expansive gesture backwards with his leg turning to face the line of seated dancers, who, although they have not stirred, remain in an intensive movement of pre-acceleration. Curving over towards them and swinging his arm up and out in a large circular gesture before shooting a taut leg out towards them, his foot slaps the floor noisily as it lands. The more the other dancers ignore him, the more the hyperbole of his gestures appears to increase. In the limit or interstice between Matthew's gestures and the other dancers we see an actual present image and a virtual past image simultaneously. Any dichotomy that existed between past and present has been unthought. Time is no longer linear, it can be experienced instead as duration. The visual movement images, provoked by the dancing, have allowed actual movement to be thought of as expressing multiple temporal signs.

Finally, Matthew bends over and gathers his body in underneath him as if to contain his energy. He then breaks out, kicking again in the direction of the dancers, still motionless with their backs to him, but full of potential inherent in their pre-acceleration. As his foot slaps the ground, he also slaps the back of his wrist with his other hand. Whilst balancing on one leg, with the other stretched out behind, Matthew slaps his wrist again and finally Stine and Yalckun begin to get up. Yalckun, using exactly the same pattern of movements that Matthew did, balances on his feet and one hand leaving his other arm swinging. This adoption of moves appears initially sympathetic, but then, as Sue and Lisa also imitate this gesture, it seems more like a ripple of laughter wordlessly moving within the group. In Matthew's impetuous attempts to assert his 'natural' power over the other dancers – he is the only white male in the piece – we experience Deleuze's hallucinations: 'that we were perceiving in folds means that we have been grasping figures through the haze of dust' (1990: 4).

Matthew's authoritative gestures have lost their significance, no-one notices them or him, they have become 'dust' – fragments of forgetfulness that now no longer recognise or respect the difference between masculine and feminine. They exist as habits or refrains that recur, and if we see them as 'hallucinations' in Deleuze's terms, that fold back as recollections, then perhaps we can begin to understand the being of time as a multiplicity of lived durations, or as 'an infinity of actual fluxes' (Deleuze 1991: 82). This is because of the irrational cut which, as Deleuze indicates, means that the 'visual image' – Matthew's dancing, which, when relinking to the dancing of the others, 'develops a whole aesthetic power which reveals the layers of history and political struggles on which it is built' (1989: 256).

It is in this sense that *Shiver Rococo* provides examples of the irrational cut, of the 'noosign: an image which goes beyond itself towards something which can only be thought' (Deleuze 1989: 334). To reiterate, the irrational cut is not a second moment which follows a first, like a line, but an incommensurable 'limit or interstice' (ibid.: 278) between non-linked but relinked images. Matthew's dancings relink to the dancings of the group creating a 'noosign' or an irrational cut, parodying traditional gender roles, whilst also foregrounding the importance of time coming before space. 'What has now become direct is a time-image for itself, with its two dissymmetric, non-totalizable sides' (Deleuze 1989: 261). The irrational cut has resulted in 'a struggle marking the itinerary of a world which emerges from one historical moment to enter into another' (ibid.: 248). The danced codes of *Shiver Rococo* are thus concerned with re-linking new asymmetric experiences of time and space. They are fluid, fluctuating and multiple, in states of continuous variation, exhibiting properties that are 'objectile', a notion that Deleuze introduces in his book, *The Fold* (1993).

Objects becoming 'objectile'

In his book *Difference and Repetition* (1994), Deleuze's asymmetric synthesis of the sensible suggests that exchange between two *kinds* of actualising multiplicity – one chiefly quantitative and discrete and one chiefly qualitative and continuous – gives rise to different kinds of repetition.[6] We can see this in the dancing of *Shiver Rococo* when different dancers are repeating the 'same' movement material, but inevitably differently, because they inflect it with different affects and percepts. In this sense the 'objects' of dancing become 'objectile', a term Deleuze introduced to describe the new consistency of folded objects liberated from the idea of a mould promoting individual style (1993: xix). The object is no longer confined to the 'idea of a standard that still upheld a semblance of essence and imposed a law of constancy ("the object produced by and for the masses")' (ibid.: 19). Time replaces space, since 'the object acquires a new status when it refers no longer to a spatial conception of molding, but a "temporal modulation" or a "continuous variation of matter"' (ibid.: xix), or folding. When objects of dancing become 'objectile' in this way, they are not static and rock-like, but have a consistency. Or, in other words, as José Gil claims: 'there is no single visual or kinaesthetic image of the dancing body, but a multiplicity of virtual images produced by movement that mark so many *points of contemplation* from which the body perceives itself' (2006: 24, his emphasis).

The danced text of *Shiver Rococo* consisted of a stream of irregular images of this kind. Like the 'irrational cut', they allow us to think of time images as movement images where movement is not pre-given in extension. For example, at one point Sue is arched back over Yalckun's arms. Stine, looking ahead and away from her, *bourrées* past her and pulls her up and out by the hand into a hunched forward grotesque posture. Sue's bent knees and feet are turned out, and her back, shoulders, arms and head are curved over forward. This is a bizarre juxtaposition of Stine's svelte, upright, classical body, aloof, elegant and contained, above, yet joined with, Sue, whose body has become unwieldy and monstrous, ogress-like. It is an 'irregular image', and especially so in the languages of immanence and transcendence which underwrite these danced languages of sexed difference. A *bourrée* is traditionally completed with 'finished' points in the hands and feet, but Stine kept her hands and feet moving in tiny fluctuations. The force of Sue's movements, however, was located in a shift in body weight from a centrifugal to a centripetal momentum. This set up another irregular image, disrupting and shifting norms and expectations associated with dancing bodies. Sue and Stine then split off in different directions and Yalckun bends forward and over in a similarly grotesque pose. The repetition is different because, with Yalckun's inflection of different affects and percepts, a new consistency has been folded into the danced image. It is a fluid consistency because of its continuous variation of matter. The irregularity of the images – Stine's fluctuating *bourrée* and Sue's centripetal momentum of her ogress-like form – together with Yalckun's productive repetition, are objectile. They are evidence of a 'fluctuation of the norm replac[ing] the permanence of a law' (Deleuze 1993: 19). As Deleuze indicates 'the new status of the object no longer refers its condition to a spatial mold – in other words, to a relation of form-matter – but to a temporal modulation that implies as much the beginnings of a continuous variation of matter as a continuous development of form' (ibid.).

For Bergson, the being of time as multiplicity means that it is an error to conflate the material extensions of specific temporal/spatial schema with time. Rather problems should be stated in terms of time before space. Partial temporal/spatial schema, for example the flesh of the dancing body, are relatively slow ways of 'cutting out' sense from the nonsensical overload of the immediate datum of experience, but whose lived action must be viewed as a flux amongst other fluxes. For Deleuze, the being of time is a virtual *multiplicity* in excess of all the different kinds of actualised series. As a 'difference without a concept', it has no image but is the condition of images as such. The question is always one of making disparate things dance together, of making consistency out of chaos, which is also evident in the danced text of *Shiver Rococo*.

As Stine and Yalckun pass each other, they briefly clasp hands, then Sue runs to a corner followed by Yalckun. Stine joins Matthew and Lisa, who have been standing in *relevé* feet astride and with arms out to the side. Stine and Lisa then support Matthew between them as he leans back and forward across their arms, balancing on one leg. At the same time Yalckun swings his leg over Sue and grasps her hand briefly as she rises before running into the space to join Stine who has left Matthew and Lisa. These dancing bodies are continually creating and recreating vectors in space, that are 'transpositional: a moving-through points' (Massumi 2002: 185). They are hovering in the gaps or intervals between perception and action, creating affects. Two same-sex couples emerge: Stine and Sue, and Matthew and Yalckun. They travel around the periphery, Stine and Matthew stretched upright, leading Sue and Yalckun, who have resumed Sue's earlier hunched grotesque pose. The partnerships skirt round the edge of the space, leaving Lisa standing alone in the centre. Thus, in one or two minutes, the five dancers have moved through a range of different spatial relations, seemingly at random. They have created a flux amongst other fluxes. Perceiving them in this way means thinking in terms of actualising multiplicities, of the plays within intervals giving us new experiences of space and time. These 'variable vectors' of the dance, continually in movement, can also be seen as objects that have become 'objectile' in Deleuze's terms; they have assumed a 'place in a continuum by variation' (1993: 19). We want to propose that these objectile conditions, these continua in variation, allow us to think of time images made by the singular intensities of the existential refrains as a way of reading group aggregates in dance as fluid, complex and ever-changing – on the bottomless ground of movement. Conceiving of time before space, allows us to apprehend spatial images of aggregates differently. We see them in an objectile or fluid sense. This also allows us to see movement in terms of intervals, which are both temporal and spatial. Seeing the intervals/gaps/interstices/fissures in these dancings aids us in understanding time and space durationally and differently.

The theoretical and corporeal fissures transversing the group aggregate of the five dancers appeared to emanate from between dancers creating an increasingly complex symbiosis of perspectives. This occurred between two, three, four, five dancing bodies, more if one includes the affective infolding of the audience and the writing team. Claid's use of a space surrounded by an audience on four sides meant that there was no fixed viewpoint from which to analyse the extension of pleats of matter and the ramifications between partial series of proximity and distance. The notion of 'outside' itself could only be posited in relation to variable vectors themselves in movement. Disparate dancers who were differently gendered, coloured, aged and sized, employing disparate movement vocabularies and styles, evident in the dancers' individual repeated habits or refrains, danced together in different combinations (solos, duets, trios, and so on). These continually changed, and were seen from different angles. They were also perceived differently by different audience members and writers. The aggregate of these continua, the becoming dancing, made consistency out of chaos (see, for example, Figure 4). Thus the totality of the past, because of the limitations and nature of perception, gains a partial virtual ontology in the many images cut out of it. To reiterate, the 'objects' – the production of the dancing – became objectile 'where fluctuation of the norm replaces the permanence of a law' (Deleuze 1993: 19). Irregular images flashed up between bodies 'with time enough to be seen for an instant' (Deleuze 1990: 4), but which could not have been produced anywhere else than in the sum of group relations.

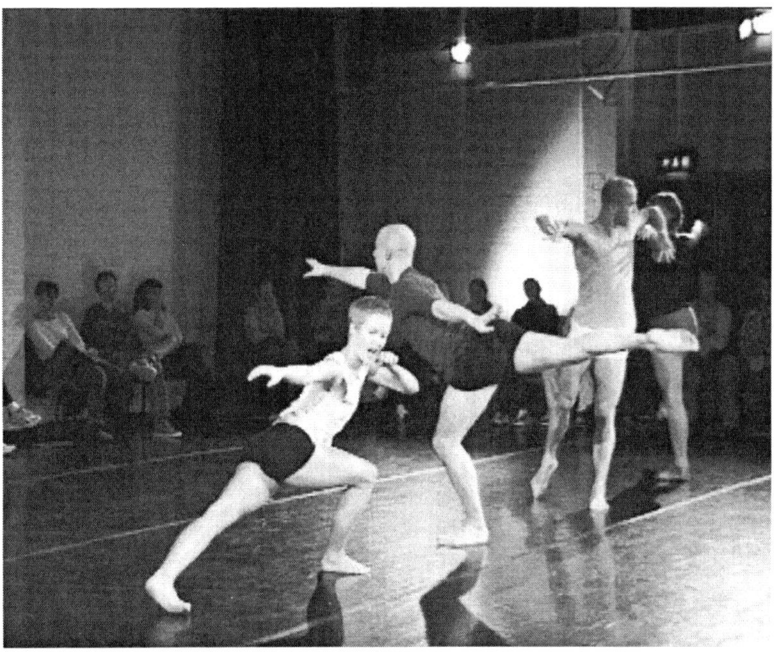

Figure 4 Lisa Turon, Matthew Hawkins, Yalckun Abdurehim and Stine Nielson in *Shiver Rococo*

Moreover, the framing of the 'any-space-whatever' of dance takes place through the techniques unique to a choreographer to make space take on a new value. Importantly, the images made by performing bodies which frame the 'any-space' with a corporeal, visceral and often visual logic and, in very embodied ways, mean that choreographic space never arrives empty. Dance creates space. It colours and vibrates it. This means considering:

> the body as a meta-phenomenon, simultaneously visible and virtual, a cluster of forces, a transformer of space and time, both emitter of signs and trans-semiotic [...] a body inhabited by – and inhabiting – other bodies and other minds, a body existing at the same time at the opening toward the world provided by language and sensorial contact [...] a body that opens and shuts, a body that endlessly connects with other bodies and elements.
>
> (Gil 2006: 28; see, for example, Figure 5)

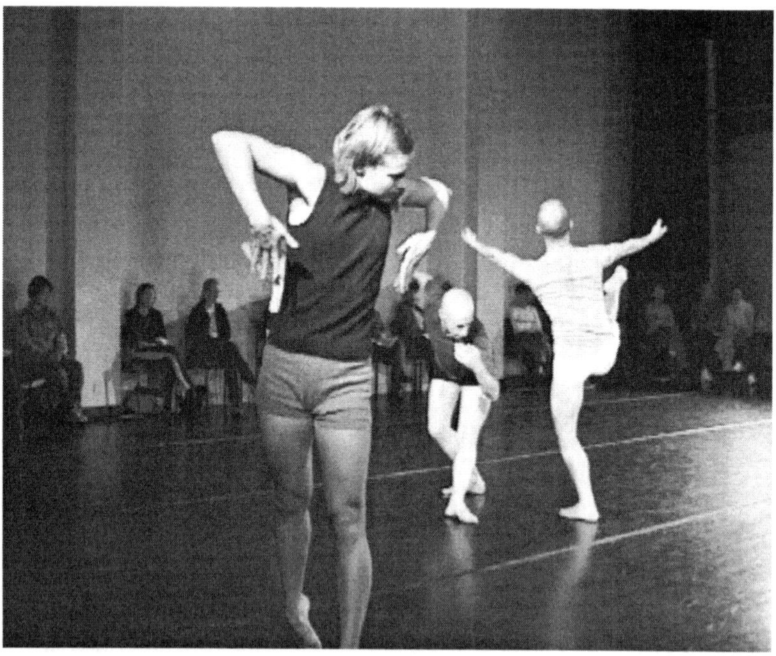

Figure 5 Stine Nielson, Matthew Hawkins and Yalckun Abdurehim in *Shiver Rococo*

A dancing body open to the world in this way is capable of 'worlding', 'always involved in a reciprocal reaching-toward that in-gathers the world even as it worlds' (Manning 2008). It is this openness of the dancing body we want to align with Bergson's affective openness.

Bergson's affective openness

Bergson's affective openness, like the openness within the continual movement of the dancing, supplies philosophy with the intuitive impulse it needs to turn and return beyond the relative stasis of mythic closures. In *Matter and Memory* (1988), Bergson argues that there are no things in themselves but movement between states, and that there is an adaptive interaction between two kinds of intelligence within human consciousness. One is speculative and shapes matter in extended space. It must depend upon the necessary illusion of one duration to give it a sense of undivided experience, and it corresponds to the quantitative arm of analysis in Deleuze's asymmetry. The second kind of intelligence is comprised of a multiplicity of qualitative durational registers, which in Deleuze's *The Fold* (1993) are more or less aligned to Leibniz's world of animal perception under the five senses of the baroque house. This intelligence is wordlessly apprehended through intuitive affects.

Bergson never tells us where images come from, the mind is an image amongst other images, all of which have a duration, or partial lived time specific to them. Instead of a split in the subject, Deleuze's reading suggests that 'time has to split at the same time as it sets itself out or unrolls itself: it splits in two dissymmetrical jets, one of which makes all the present pass on, while the other preserves all the past' (Deleuze 1989: 81). The perpetual foundation of time is 'splitting [that] never goes right to the end' (ibid.). Bergson also posits no fundamental difference in kind between the materiality of the mind and the material composites, which can and cannot be touched upon, and he places the body in an open network with the mobile material world: 'my body is, then, in the aggregate of the material world, an image which acts like other images, receiving and giving back movement' (Bergson 1988: 19). This embodied materialisation of the mind where it is likened to the body, placing the body/mind composite in an open network with the mobile material world, is evident in much of the *Shiver Rococo* danced text. This occurs as space is fragmented and splintered by the continually changing dynamics of the dancers' multiple gazes, and because audience members could not maintain 'oneness' or singularity by gazing from a single perspective. This disorientation is further underlined in the choreography. The dancers shift and *move through* a range of different angles. Their bodily centres of gravity are continually being displaced and moved off-centre. Although there are times when one, two, three, four or all five dancers appear still, because of their distorted and disorientated bodies there is a sense in which they are never still or fixed in a rooted sense. They are, rather, in the state mentioned earlier of 'pre-acceleration' or 'incipient action'. They are 'receiving and giving back movement' in Bergson's terms, open rather than closed, illustrating the perpetual foundation of time which is 'splitting [that] never goes right to the end' (Deleuze 1989: 81). Different parts of their bodies point in different directions. There is little sense of unity or complacency in the body. Much more, there is a sense of ever alert energy darting in different directions. These are movements of the not-yet, creating space vibrating with potential action. In *Shiver Rococo*, the intricacies of twists, turns, curls, folds and undulations in the rococo are shivered throughout the performance, and shivering, often associated with affects which go beyond the limits of speech, is unsettling.

One example of the openness of mind and body to the mobile material world is a recurring sequence in which a dancer's head is placed on another's stomach. This irregular interconnection between bodies happens four times. The first comes when Matthew strides across to Lisa. He is grotesquely bent double with his head down. As his head touches Lisa's stomach, he reaches his arm around her waist. She immediately turns, slaps her hip and looks away aloof, raising her arm in the air. She takes Matthew's head off her stomach with her other hand and lifts it up whilst walking around him to face him again. This time he raises both hands, slowly places them over her ears and lowers *her* head to *his* stomach. He holds it there for a few seconds before releasing it and she turns away. As Manning suggests, individuation 'also becomes infraindividuation' which can be involved in creating:

> a vocabulary for how movement becomes thought and vice-versa. In this eternal return of movement-becoming-thought and thought-becoming-movement [...] movement tells stories quite differently than does a more linear and stable historicisation.
>
> (Manning 2008: n.p.)

Much later, a similar conjunction occurs, preceded by Lisa, bent over double and stepping backwards such that her curved back pushes Matthew's outstretched leg around a quarter turn. Simultaneously, Stine moves under Matthew's arm, lifting him off the ground with her shoulder. The two women supporting Matthew take him around in a full turn and lower him towards the ground. He is then held upright from behind by Stine's hands on his waist, whilst he takes Lisa's head in *his* hands placing it on his stomach, repeating the earlier move exactly. This is like a body receiving and giving back movement in Bergson's open network with the material world, restoring what it receives. Between these two grotesque moments, there is another when Matthew strides backwards, whilst bent over forwards with Yalckun's head in his stomach, pushing Matthew backwards. This is not mimicry of animals but a human, laughing, dancing body mocking a logic that has asserted that the material flesh should be denigrated for being a kind of animal with micro-perceptions, 'pricklings' at work in its macro-perceptions.[7] These bizarre, partial couplings in *Shiver Rococo*, in their ambiguous plays with bodily connections – the head, conventionally associated with the brain or mind, contacting the stomach, associated both with the micro-perceptions of nourishment and, in women, with the womb as a provider of nourishment – graphically bring the materiality of mind and body together in an affective openness with the world. Furthermore the (ironic) reversals of male/female relationships (have the potential to) go beyond gender.

Bergson argues that no member of a species is ever exactly the same as another, either in different points of time or different points of space. It is the creative synthesis of the 'certain limits', the singular, partial acts of recollection in symbiosis with the future act of perception which produces the 'consistency' of the recognisable virtual image. There is, therefore, always something new in recognition which moves it beyond the vicious circle of hermeneutics. For Bergson, there is nothing remarkable in the fact that perception takes place; it is the acts of perceptual subtraction and multiplication which need to be explained. While Bergson fails to provide satisfactory theories concerning the economies at work in these durational cut outs, what he does do is neatly debunk the notion that the successions of any one durational image, or even two, could capture the movement of time itself. As the movements of dancers continually demonstrate, duration is more complex than this. As with the time image, what is intuitively apprehended in these dancing unfolding and refolding images is not chronological time, but 'the perpetual foundation of time' spatially expressed as a doubling of the actual and virtual within irregular movement images.

Shiver Rococo ends with Lisa passing her hand over her eyes closing them. As Yalckun approaches, she reaches out into space, passing her hand over his eyes, closing them too. Each of them is blinded. They then improvise together, sensing each other kinaesthetically, haptically, through touch. Openness is key. This openness is affective, full of vibratory potential. The two dancers inhabit the same space. Their sensing connects them, through the micro-conceptions or pricklings of acute awareness that minutely fold and refold and unravel in every direction between them (Deleuze 1993: 86). Although they are continuously in contact, it is not their touch *per se* that holds them together, but the 'act of *reaching toward*' which occurs through the bodies' relationality, through their worlding (Manning 2007: 5). They support each other, extending limbs into space, one body lifting another off the floor, miraculously soaring through the air, creating a continually moving sculptural form. Bodies stretching and arching, folding and refolding one within and around another (see Figure 6). Legs, arms and torsos of the two bodies intertwine making it difficult to tell one body from the other. Throughout, the dancers, through their affective openness, are creating and extending space towards and around each other, unfolding the experience of space as they move. Guided entirely by the sense of touch rather than vision, they are both subject and

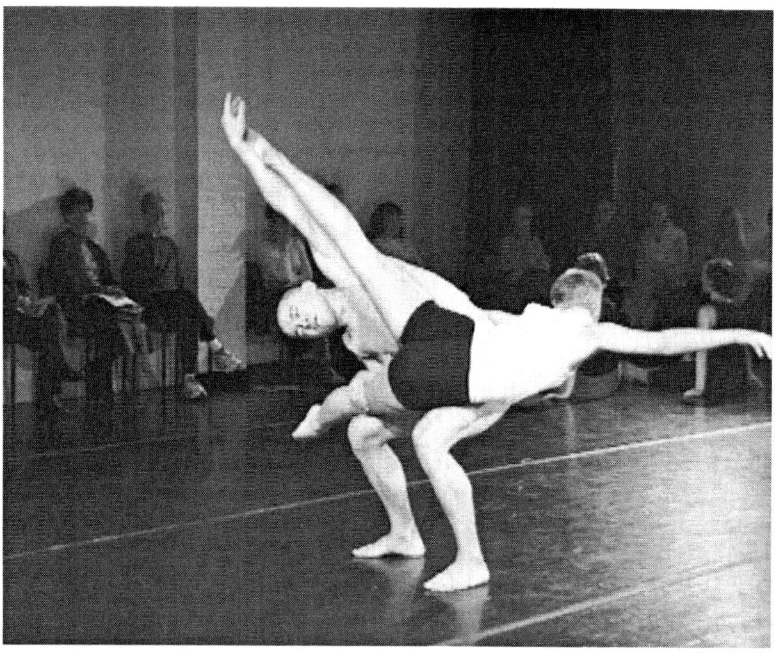

Figure 6 Yalckun Abdurehim and Lisa Turon in *Shiver Rococo*

object. They create an enfleshed space, an embodied space of touchings and sensations that do not depend on the look or gaze. The pair, by losing vision or sight in an inner space of reflection, which is the space of performance and of the dancing for them, are discovering for themselves and the audience a new becoming sense of spacing, of space-time, of time.

It is fitting that *Shiver Rococo* concludes in this way since this 'blind duet' most affectively and effectively embodies the sensing, reaching-toward, worlding and becoming that has infused the dancing bodies throughout and that has allowed us to begin to re-think temporality. In their reaching toward each other, these dancing bodies are in excess of themselves. They 'engage symbiotically, incorporeally, virtually, always becoming more-than they already seem to be' (ibid.: 13). They are moving 'toward something that is not-yet'; shifting 'toward the virtual', and by reaching toward, they are engaging 'with what Deleuze calls the crystallization of time where the actual and virtual coincide' (Manning 2007: 11).

Conclusion

No one group aggregate could hope to produce a mnemotechnics adequate to the multiple durations of just one moving body. Bodies remain, as Deleuze puts it, 'badly analysed composites' (1991: 18). The openness of the movements and choreography of *Shiver Rococo,* because they involve a give and take of movement, allow us to align them with Bergson's ideas of affective openness. All the senses are involved in dance. The embodied ambiguities suggested by *Shiver Rococo*'s choreographical play between the affective extremes within visual, spatial and temporal referents and the hapticity of movement languages have been opened up with the assistance of the movements of thought in ambiguities between Deleuze and Bergson. Dance works towards replicable gestures in its story-telling function and creates a being of sensation, but the difference between this being of sensation and other art forms is that it is a 'noosign', an irrational cut, always outside the houses of its images. It cannot, as its condition of being, be reproduced identically. The movement only ever occurs once, interrupting all sensory architectures, which are the makers of its monuments. If movement is stated in terms of time rather than space, as we have explored here through irrational cuts, objectiles and affective openness, then any movement motif is an aggregate of the multiplicity of temporal fluxes, which make up each dancer's corporeal signature.

With all the problems of writing endemic to our practice, the movements of *Shiver Rococo* remain in excess of the monuments we have made of them here. According to Claid's description of her praxis, there is a creative gap between the thinking of dance and the production of these works of art, which is embodied and best reflected on after the event. This silent uncertain space is a place where problems of time and space can be negotiated. As much as dance creates varieties of movement which are also becoming, its chosen medium is, more than the other arts, a zone of indistinction through which its being of sensation appears.[8] Choreography is neither chaos, pure becoming, nor the historically conflated imperceptibility of the private sphere. Compared with other art forms, it takes a deeper (quieter) durational layer of continuous variation as its plane of composition. As a result it provides an apt arena for reconfiguring time/space.

To rethink temporality is necessarily also to rethink the dancing body as a sensing, becoming body. This sensing, becoming body, not only open to but also reaching toward others, dances throughout this book. By reaching toward others it challenges the Cartesian notion of a rational unified subject and points towards individuated rather than individual bodies, and towards relations with others that suggest communities and politics to come.

Notes

1. See www.embamb.com for further details, also Chapter 11 of this book which focuses on the third choreographed piece of the trilogy, *Remember to Forget* (2003).
2. Ruth Chandler and a postgraduate student, Andrew Wilford, joined Valerie as writers on the pilot *Embodying Ambiguities* project in 1999 which produced *Shiver Rococo*. This pilot project led to the major *Embodying Ambiguities* research project co-directed by Briginshaw and Claid (2001–4) – see www.embamb.com. The origins of this paper reside in the joint philosophical connections Valerie and Ruth made 10 years ago. The essay that forms this chapter has taken a dance trajectory with Valerie as the major author. Ruth is also working on a forthcoming essay, growing out of the original paper, to be published elsewhere, that takes a more philosophical trajectory, where Valerie will be the minor author. Working on these two new Deleuzian assemblages together is an example of the kind of productive collaboration on which this book is based.
3. By 'rethinking temporality' we do not mean rethinking time, since the perpetual foundation of time is the condition of thinking, it cannot be rethought. We are concerned rather with how the extended repetitions of dance, evident in its physicality, operate temporally and can suggest different ways of experiencing time. Rethinking temporality and specifically the

distinctions between time and history are the basis of Ruth Chandler's PhD project from where much of the thinking for this chapter came.

4. Writing in these conditions means that there can be no authorial 'I' behind the 'we' in use here and this universal fiction is not extended to the readers of this text. Our 'we' is a situated multiplicity and expresses the inclusive disjunction between the writers who are also readers of each other.
5. In a grammatical text this would be transition in the singular, but for dance we want to open this up to allow for the plurality of movement transitions that occur.
6. For a more detailed engagement with Deleuze's ideas in his *Difference and Repetition*, see Chapter 12 in this book.
7. The multiplicities which underpin Bergson's and Deleuze's philosophies extend to seeing all life – human, animal, plant and so on – as similarly multiple and interconnected in an open world. We are making the point here that the grotesque movement of the dancers that some may consider animal-like is not imitating animals. Instead we are suggesting that it foregrounds our interconnections in an open world, where the divisions constructed between humans and animals are false. Its performance undermines the hierarchies that elevate human life over that of animals. 'Pricklings' are, as Deleuze describes them in *The Fold*, tiny perceptions evident in 'the animal that anxiously looks about, or the soul that watches out', they are 'minute, obscure, confused perceptions that make up our macroperceptions, our conscious, clear, and distinct apperceptions', they are 'little folds that unravel in every direction, folds in folds, over folds, following folds' (Deleuze 1993: 86–7).
8. The importance of sensation for dance in Deleuzian terms is explored in more detail in Chapter 11 of this book, which focuses on the third choreographed piece of the *Embodying Ambiguities* research project trilogy, *Remember to Forget* (2003).

4
Façade, Elvis Legs, and the Humorous Pleasures of Dancing

Ramsay Burt

This chapter compares sections of two English dance works from different periods of the twentieth century that refer in humorous ways to the pleasure of dancing as popular entertainment. These are the *Tango Pasodoble* from the 1931 ballet *Façade,* choreographed by Frederick Ashton, and Lea Anderson's *Elvis Legs,* created in 1995 as part of the evening-length programme *The Featherstonehaughs Go Las Vegas.* The *Tango Pasodoble* in *Façade,* initially performed by Ashton himself and Lydia Lopokova, was played for laughs, both in its choreographed comic incidents and in the way Ashton and Lopokova were knowingly cast against type. Thus Ashton, an Englishman, was cast as a foreign professional dancer or 'gigolo', while Lopokova, a Russian ballerina, was cast as an English debutante.

The humour in *Elvis Legs* lies in its playful tribute to an iconic rock 'n' roll figure, and in the dancers' knowing enjoyment of what they are doing. It is not a comic piece but one which raises a smile as the three dancers take on the movements of the 'King' of rock 'n' roll, but perform them in a fragmented, minimalist, de-individuating unison. The dancers in both works reveal a distinctly modern detachment. They not only perform, but are also aware of the fact that they are performing, and seem to stand aloof and 'know'. The ironic distance between these different performative levels provides a space for a kind of knowing wit and humour that, I suggest, can be seen as distinctly modern.

The philosopher Simon Critchley has argued that humour is a modern phenomenon. He points out that the 3rd Earl of Shaftesbury, writing at the beginning of the eighteenth century believed that humour has to accord with taste and thus has the potential to make people aware of shared social values. Critchley therefore links the emergence of humour to the development of a democratic public sphere. Compared with the carnivalesque buffoonery of the pre-modern age, Critchley argues: 'Jokes are forms of abstraction that place in abeyance our usual modes of reaction, whether veridical or moral [...] humour lets us take up a disinterested, theoretical attitude towards the world, but it does this in an eminently practical and interesting way' (2002: 87–8). It is this disinterested attitude towards the world that marks such humour as modern.

Where the tango is concerned, this distinctly modern sense of detachment is one which Marta Savigliano has identified in her description of the passionate state through which dancers achieve what she calls a tango 'high', where 'two bodies communicate perfectly (and look perfectly and beautifully matched)' (1998: 104). She elaborates on this through the following description:

> Absentmindedly, tightly embraced, their torsos tilt towards each other in a delicate balance, their legs tracing sinuous paths on the dance floor, muscles fully alert to the doing and undoing of mutually provoked entanglements. Their improvised steps surprise each other, and yet the music – the rhythm and the melody, they insist – hold them together, prompting the smooth continuity of the conversation between these distinctly gendered bodies.
>
> (Savigliano 1998: 104)

The dancers, in this description, seem detached from one another and thus surprised by each other's steps, while at the same time looking perfectly and beautifully matched. Savigliano calls this: 'a paradoxical state of abandonment and full control, of bodily awareness and mental disengagement' (ibid.). What is most relevant for my argument here is the relationship between different performative levels. The male and female partners are dancing but at the same time observing what is evolving between them in a knowing, modern way. This, I shall show, takes place during the performance of the *Tango Pasodoble*.

There is no interaction between dancers in *Elvis Legs*, where the same high-energy solo material is, in effect, danced by each of the three dancers, first on their own and then in unison. Its quality of humorous detachment and irony derives from the contrast between the dancers' self-absorbed manner of self-presentation and the cultural associations of the material itself. As dance movement and gesture, it evokes a heroic, narcissistic, egotistical male ego, but this is defused by the choreography's minimalist, fragmented structure and the fact that the dancers seem isolated from one another as they dance in unison, neither looking at each other nor acknowledging the audience. In a different way from *Tango Pasodoble, Elvis Legs* therefore exudes an ironic quality of humorous, modern detachment.

Performing alternative cultural identities

These two dance pieces use humour to reveal ways in which, as embodied beings, we negotiate the social and cultural discourses through which ideas about gender and other aspects of identities are maintained. Judith Butler points out that 'the body has its invariably public dimension: constituted as a social phenomenon in the public sphere, my body is and is not mine. Given over from the start to the world of others, bearing their imprint, formed within the crucible of social life, the body is only later, and with some uncertainty, that to which I lay claim as my own' (Butler 2004: 21). Anyone who has learnt to dance knows the extent to which one's body is given over to others before it becomes one's own. Performing one's gender, Butler argues, is like performing a play: 'gender is an act which has been rehearsed, much as a script survives the particular actors who make use of it, but which requires individual actors in order to be actualised and reproduced as reality once again' (Butler 1990: 277). One cannot refuse to perform the gender that is ascribed to one, but as one lays claim to one's body, one lays claim to the means through which one responds to and interprets this demand. The possibilities for interpretation are circumscribed through the ways in which gender norms are institutionalised through cultural forms like dance. But institutional structures tend to be inefficient and can sometimes have unintended, occasionally humorous consequences, particularly where their implementation is dependent on human agency. It is the resulting gaps and inconsistencies within social identities as they are performed that are the main focus of my readings of the humour of these two English dance pieces.

Individuals are always complex and multifaceted, finding their identities within a wide range of economic, cultural and political determinations. It would be a mistake to isolate issues concerning gender from these other components of identity. What makes normative masculine identities particularly important, however, is the function they perform in maintaining an imbalance in power relations between men and women. Consequently, the underlying ideologies controlling masculine behaviour have a distorting effect on the kind of ideal that they propose, so that the quintessence of manliness becomes in effect unattainable. This is partly what I think the psychoanalytic theorist Jacques Lacan was referring to when he proposed that the penis can never attain the status of the phallus. William Washabaugh and others have argued that the tango expresses a masculine melancholia that derives from the disappointment of impoverished, slum-dwelling European migrants in Buenos Aires at the start of the twentieth century. In some accounts, members of this social group are identified as the tango's first exponents. Other writers have written that the melancholy mood of tango songs and short stories evokes a masculine fall from grace or a sense of loss of authentic masculinity. In an urban, English working-class context during the 1950s, Elvis Presley's songs were taken up by a social group – teenagers – for whom social, political, and economic circumstances had undermined traditional, 'authentic' working-class community values (see Cohen 1980). *Tango Pasodoble* and *Elvis Legs* both acknowledge in humorous ways the complex and sometimes subversive ways in which popular social dances can intervene within norms of social behaviour. This is because of the way they each create a sense of shared community among those who enjoy dancing in the particular style of that work and appreciate watching such dances performed well.

Façade and *Elvis Legs* exemplify two different stages in the English reception of globalised social dance phenomena – the tango and rock 'n' roll. The tango in *Façade* suggests that an Argentine melancholia has perhaps translated, in this dance piece at least, into a distinctly English sense of humour concerning men's disappointment. *Elvis Legs* represents a female choreographer's celebration of a kind of male dancing she enjoys, capturing its energy and redirecting it in ways that defuse its aggression and thus widen its appeal. Part of the piece's humour derives from the subversive nature of doing so. My proposal is that a modernist detachment is inherent in the humour of these pieces' appropriation of aspects of these dance styles as they were received in Britain. By articulating this detachment, the *Tango Pasodoble* and *Elvis Legs* not only draw their beholders' attention to the existence of cultural identities but also reveal the gaps and inconsistencies within the norms that govern them. Their humour gently hints at the lenient way in which society excuses most men for failing to live up to an unattainable ideal. This is humorous because, as Simon Critchley observes: 'humour consists in laughing at oneself, in finding oneself ridiculous, and such humour is not depressing, but on the contrary gives us a sense of emancipation, consolidation and child-like elevation' (2002: 95). This, I suggest, is what *Tango Pasodoble* and *Elvis Legs* each achieve, in different ways, through their use of what Critchley calls an 'ever-divided humorous self-revelation' (ibid.: 86). Such self-revelations allow these two pieces to celebrate social dancing as an area in which it is sometimes possible to create a space for the performance of alternative identities.

Tango Pasodoble

Ashton's *Façade* was set to an orchestral suite by William Walton that had initially accompanied a cycle of poems by Edith Sitwell. The ballet was originally commissioned by the Camargo Society – a subscription society that annually, from 1930 to 1933, hired a London theatre for a short season, commissioned new ballets and hired dancers and musicians to perform them.[1] Two of the key players in the society's affairs were the economist John Maynard Keynes, its treasurer, and his wife Lydia Lopokova, who had been one of the principals of Diaghilev's Ballets Russes.[2] Both were members of the Bloomsbury Group, although, as Judith Mackrell notes, Lopokova was not entirely accepted by some of its members. The Bloomsbury Group consisted of writers, painters and art theorists, including Lytton Strachey, Virginia and Leonard Woolf, Vanessa and Clive Bell, Duncan Grant, E. M. Forster and Roger Fry, most of whom lived in or around Bloomsbury Square in London. As sociologist Janet Wolff (1996) has argued, the Bloomsbury Group used their connoisseurship of modernist art to reassert class privilege and compensate for their potential loss of social and economic status. They were in danger of losing this as a result of the erosion of class distinctions arising from the impact of commercial and industrial modernisation at the start of the twentieth century. In the early 1920s, ballet had become an area through which Bloomsbury connoisseurship was exercised because of the way Diaghilev had brought ballet together with modernist painting, poetry and music.[3] The Camargo Society was a response to the vacuum left by Diaghilev's death in 1929, putting on programmes that catered for the London ballet audience which Diaghilev had created. *Façade* brought together ballet, modernist poetry, music and painting in a way that Diaghilev had pioneered.

Edith Sitwell wrote and published the poems that eventually became *Façade* between 1918 and 1922. Her brothers Osbert and Sacheverel had seen Diaghilev's ballet *Parade,* a collaboration between Picasso, Cocteau, Satie and Massine, in Paris in 1917. They encouraged their sister to conceive of *Façade* as a consciously avant-garde musical entertainment. It thus became a vehicle through with to launch their young protégé, the composer William Walton. The Sitwells, like the Bloomsbury Group, constituted a tight, closed circle who used connoisseurship of modernist art in similar ways to secure class privilege.

The resulting piece was first performed in 1923, with Edith Sitwell herself reciting or intoning (but not singing) her poems to an orchestral accompaniment.

Wilfred Mellers and Rupert Hildyard (1995) have argued that the 'Englishness' of Sitwell and Walton's *Façade* is produced through articulating a historically specific, anti-bourgeois nostalgia for a lost Edwardian golden age. It evoked the disappearing world of popular seaside entertainment in the late Victorian and Edwardian period which Edith Sitwell remembered from childhood stays in her family's seaside holiday home in Scarborough (see Sitwell 1987). The poem *Tango Pasodoble* seems to refer to these holidays. It begins:

> When
>   Don
> Pasquito arrived at the seaside
> Where the donkey's hide tide brayed, he
> Saw the banditto Jo in a black cape
> Whose slack shape waved like the sea

Frank Howes notes that Walton's setting of this poem includes musical references to the popular song 'I do like to be beside the seaside' (1965: 16).

The ballet *Façade* received its premier as part of a Camargo Society programme on 26 April 1931 at the Cambridge Theatre in London. It was set to an orchestral suite of Walton's music performed on its own without the poems. It was danced by Lopokova together with Ashton, Alicia Markova and some dancers on loan to the Camargo Society from Marie Rambert's Ballet Club. Rambert immediately bought the décor and costumes and it was first danced at the Mercury Theatre by her company a few weeks later.[4] It consisted of seven numbers:

1. Scotch Rhapsody – pas de trios: Prudence Hyman, Maud Lloyd, Antony Tudor.
2. Yodelling Song – a milking scene à quatre: Lydia Lopokova, Frederick Ashton, William Chappell, Walter Gore.
3. Polka – solo en pointes: Alicia Markova.
4. Valse – à quatre danseuses: Pearl Argyle, Diana Gould, Lloyd, and Hyman.
5. Popular Song – à deux hommes: Chappell, and Gore.
6. Tango Pasodoble – à deux: Ashton, and Lopokova.
7. Finale, Tarantella Sevillana – ensemble: the company.

There were several reviews of this programme. The *Daily Mail* thought *Façade* was 'full of parody and fun', the *Daily Sketch* found it 'full of quaint humour'. The *Daily Telegraph* did not mention *Façade* at all but judged that there was a wealth of native talent: 'Altogether it is evident that we have the dancers, the choreographer (Frederick Ashton is a pillar of strength) and the musicians for an indigenous theatre of dance.' The sub-text here is the fact that, up until then, there had not been an English ballet tradition as such. *The Times* reviewer mildly approved of *Façade*, observing that the ballet took nothing from Sitwell's poems but admirably translated the spirit of the music. Stating that the 'Yodelling Song', the Polka and the Tango were particularly successful, it ended by remarking: 'One doubt arises. If the Ballet is going to laugh at itself so freely, it must take care in future not to laugh in the wrong place.'[5] The reviewer, like the Earl of Shaftesbury 200 years earlier, evidently believed that humour must accord with taste. The *Manchester Guardian* considered *Façade* to be the principal novelty of the evening and particularly appreciated Lopokova's performance in the *Tango Pasodoble*. 'No other dancer has ever been able to assume for comic ends an expression quite so vacant, like a doll in a dream, and to belie it so richly by the subtlety of her dancing.'[6] This exemplifies the kind of modern irony I discussed earlier.

As I have already noted, *Façade* is full of reference to foreignness. The décor designed by the modernist English painter John Armstrong included a Swiss chalet with a Dutch door. Like the décor, the ballet had a *faux-naive* exoticism. There was an Alpine yodelling song and a *Scotch* (not Scottish) *Rhapsody*. Two of Sitwell's titles mixed countries up: the *Tarantella Sevillana* and the *Tango Pasodoble*. A Tarantella is Sicilian rather than from Seville, and while the tango is from Argentina, a pasodoble is a march at a Spanish bull fight. Sitwell was playing with the idea that the English upper-middle classes liked to feign ignorance about anything foreign.

Ashton had learnt the tango while living in Lima as a child and young man. In the *Tango Pasodoble*, he portrayed an oily 'dago' who is a professional dancer. 'Dago' was a pejorative term, now largely defunct, for someone of Italian or Mediterranean appearance. Dorothy L. Sayers' 1932 detective novel *Have His Carcase*, one of her series about Lord Peter Wimsey and his fiancée Harriet Vane, includes the following description of a professional dancer at a provincial seaside hotel:

> Miss Harriet Vane, in a claret-coloured frock, swayed round the dance-lounge of the Hotel Resplendent in the arms of Mr Antoine, the fair-haired gigolo.

> 'I'm afraid I am not a very good dancer,' she remarked, apologetically.
>
> Mr Antoine, who was rather surprisingly, neither Jew nor South-American dago, nor Central European mongrel but French, clasped her a very little more firmly in his competent professional arm.
>
> (1977: 83)

It is not exactly clear whose xenophobic, anti-Semitic prejudices are being voiced here – those of Vane, the author, or those of upper-middle-class English society between the wars, but gigolos were clearly objects of distaste even if Mr Antoine was meant to be less objectionable than most. I like to think they may have been dancing a tango, because Harriet Vane enjoys the firm but open way in which he is leading her. Belinda Quirey explains the English social dance context for this:

> In the 1920s the change to a totally new deportment and style [of social dancing] left many middle-aged men in the world of the waltz, and unwilling to learn new tricks. Many of their middle-aged wives, on the other hand, had taken to the new style. There was therefore a demand for good partners and this was supplied by many professionals. As a rule these were very pleasant men and nearly all the ones I knew I liked very much. But to the normal English male they were simply gigolos. 'Damn dagos', was the usual condemnation.
>
> (Quirey 1976: 87)

There is a complicated in-joke involved in the *Tango Pasodoble*: Ashton as an Englishman (though born and brought up in Peru) impersonates a 'dago', and Lopokova, Russian (but married into the English upper-middle class), impersonates an English debutante. A debutante was a young upper-class woman being introduced into London's elite social circles for the first time. *Façade*'s debutante was awkwardly naive and dressed in a comically ill-fitting ball gown made of cheap, tasteless materials. Audiences at the time would have interpreted this lack of taste as a sign that she wasn't quite one of their class. Alicia Markova, who danced the Polka in the original production but subsequently also took over the debutante's role, later remembered:

> The whole thing between Sir Fred and the debutante was almost as if he didn't wish to touch her. The debutante is naive; everything he tries to teach her goes wrong. There were roars of laughter all the way through. Just as she thinks she is getting the rhythm of the walk

across in front of him, he tips her over in the air. With a fast lift, where he whirls her round on the ground and they come face to face, again there were roars of laughter.

(Markova in Jordan and Grau 1996: 177)

As *The Manchester Guardian*'s critic had recognised, a dancer requires a high level of technical skill and performance experience in order to convincingly appear to be a bad performer. For the beholder, such good bad performance demands split levels of appreciation, one straightforward, the other ironic and humorous. Because performance of the tango involves a degree of detachment, *Façade* asked the viewer to appreciate the *Tango Pasodoble* in an ironic way. One can detect this disinterested quality within the piece's dramaturgy as Markova describes it. The Dago doesn't want to have anything to do with the Debutante, while she is surprised by what he does, just as Savigliano notes that in the tango each partner's improvised steps surprises the other.

Ashton's humour in the *Tango Pasodoble* had two targets. Rather than laughing at social dancing in general, one target was those who were naive and incompetent at it. This joke was shared by those who were knowledgeable and enjoyed dancing. Their laughter created a bond with fellow audience members and performers who were passionate about the tango. This was, nevertheless, a bond between members of a particular fraction of the English upper-middle class. The other target, as I have already pointed out, were foreigners, but its witty attack on them was complicated by questions about manliness. On the one hand, Englishness seemed to be identified with the normal and unremarkable by presenting foreign dancing as exaggerated, colourful and remarkable. But, in the *Tango Pasodoble*, this insular worldview seemed to be upended with the South American-born English dancer cast as a foreigner and the Russian ballerina cast as a young English girl. The dominant masculine role of the male tango dancer, particularly that famously exemplified by Rudolf Valentino in *The Four Horsemen of the Apocalypse*, is in line with normative ideals of manliness. Read in an English context, however, this gender performance appears to be foreign. Because this foreignness is a source of amusement, the *Tango Pasodoble* brought into play an ironic distance that excused English men for failing to live up to what the piece suggested were, after all, unattainable, manly ideals.

Elvis Legs

Just as *Façade* became a more popular and enduring ballet than its initial reception indicated, *Elvis Legs* became better known than the show for which it was initially choreographed. Following the initial production of *The Featherstonehaughs Go Las Vegas* in 1995, *Elvis Legs* was revived in 2000 on its own for performance during a *Ducky Night*, a gay cabaret night, at the Vauxhall Tavern in London, and revived again in 2003, this time by the Cholmondeleys, Lea Anderson's female dance company, as part of *Double Take*, her contribution to the programme for the official opening of the new Laban building. (For three weeks I watched *Elvis Legs* in rehearsals for this.) In 2004–5 *Double Take* was extended and developed into a new, evening-length programme which toured England celebrating the Cholmondeleys's twentieth anniversary.[7]

The Featherstonehaughs Go Las Vegas was a pretext for stringing together scenes loosely based on the idea of Las Vegas that included gamblers, boxers, gangsters, drunks and famous entertainers like Elvis Presley. Presley has become an iconic figure on a global and intergenerational scale. The Elvis of *Elvis Legs* is seen both in relation to US youth culture and from an idiosyncratically British perspective that draws heavily on the British reception of rock 'n' roll music in the 1950s.

Elvis Presley's extraordinary early success in the United States exploited the fact that, as a poor, Southern white performer who synthesised both black and white music and dance traditions, he was able to appeal to a huge white American record-buying public. As record executive Joe Bihari recently recollected: 'There was very little airplay for R&B record, because there were very few black DJs, and very few black stations that played black music. Also, it wasn't easy, particularly in the South, for white kids to bring home black music in the 1940s. Their parents frowned on it' (quoted in Petridis 2004: 4). In 1954, Presley's *That's Alright Mama* and Bill Hailey's *Rock Around the Clock* were the first big rock 'n' roll hits, but there are a number of earlier songs by black artists, including Ike Turner's 1951 *Rocket 88* and Roy Brown's 1948 *Good Rocking Tonight*, that could compete for the status of the first real rock 'n' roll record (ibid.).

It is necessary to acknowledge the privilege which Presley enjoyed as a white performer in being able to exploit black musical and performance styles. But one should also, at the same time, recognise rock 'n' roll as a hybrid rather than a purely black phenomenon, and note that Presley's dancing was one area in which his difference from black performers can be clearly identified. I refer here to what I shall call his partial failure to dance in a black way. As Martin Hargreaves (1998) has observed, Presley's earliest widely broadcast television performances became subject to controversy and censorship because of the way his rhythmic body movements focused attention towards his pelvis. Through association with the imagined hypersexuality of African American men,

'Elvis the Pelvis' was condemned as a threat to the moral codes of normative white masculine behaviour. But, as Hargreaves has pointed out, there is a tightness and awkwardness about Presley's pelvic movements that contrasts with the more fluid and relaxed whole-body movements of comparable black performers like Chuck Berry and Little Richards.

It is this distinctive, individual and rather quirky dancing style of the young Presley which Anderson's *Elvis Legs* has taken as its starting point. However, *Elvis Legs* also draws on the somewhat different responses that rock 'n' roll inspired in Britain in the 1950s. Cultural theorist Dick Hebdidge has suggested that, from the point of view of young English working-class men during the 1950s, the lyrics of Elvis Presley's 'Heartbreak Hotel' or Gene Vincent's 'Be Bop a Lula', describe 'in a language familiar only through the cinema [...] a distant world the appeal of which must have been considerably enhanced by its remoteness, its unapproachability' (1979: 50). Presley, as a larger than life fantasy figure, offered a charismatic appeal for young working-class English men because his own swift rise to fame had been one from a poor, white social background and, like them, he lacked verbal articulacy. Phil Cohen (1980), in his study of this subcultural group, shows how the cultivation of a highly sophisticated sense of visual style and accompanying dance skills that were inspired by Presley offered possibilities to compensate for blocked verbal facility. Within the context of post-war Britain where working-class communities were battered and marginalised by economic rationalisations, an appreciation of largely non-verbal aspects of rock 'n' roll music and style became an area within which to recover a lost sense of community.

Elvis Presley's image, taken out of its US context and transported into a very different social and economic one for a subcultural group of working-class British youths, enabled the development of a very particular and idiosyncratic set of male fantasies of narcissism and omnipotence. Forty years later, the circumstances that had made particular details within this subcultural response so meaningful, have now been largely forgotten within the much larger range of meanings and associations that the figure of 'Elvis' has now posthumously accumulated. To unpack this a little, there is Elvis the Pelvis, with his tight and slightly awkward pelvic movements, who dared to dance in his own way despite the massive disapproval from an older generation who demonised him for supposedly corrupting (white) American youth. There is also Elvis 'The King', whose rapid rise to international fame gave him opportunities to enjoy a sense of narcissism and omnipotence while performing to huge audiences of adulatory fans. The way Anderson has choreographed and staged *Elvis Legs* enables the dancers to appear both to borrow some of the coolness of the rebel and explore fantasies of narcissism and omnipotence. As rebels they practice dancing in their own way, but paradoxically they do this in uncanny, minimalist unison. Their performance also includes the kind of larger-than-life gestures with which Elvis and other rock stars wowed their audiences, but Anderson's choreography fragments this and thus undermines any potential that these movements and poses might have for narcissistic performance. It thus excuses the dancers for failing to live up to unattainable ideals, and extends a benign, humorous understanding over the scene. This encourages the beholder, along with the dancers, to enjoy the piece's witty articulation of complex overlapping layers of cultural associations.

To make the first version of *Elvis Legs*, the dancers learnt the lower-body movements of three short extracts of Presley's dancing: his performance of *Hound Dog* on the Milton Berle Show in 1955 and, from the 1957 film *Jailhouse Rock*, the title number and the song *Baby I Don't Care*. Each dancer reconstructed approximately 20 seconds of movement from one of these and then taught it to the other two. The dancers decided that the twists and turns of Presley's pelvis during these early performances were not initiated from the upper body but mostly came from the feet and, in particular, from an idiosyncratic way of pivoting from balancing on the heel to balancing on the toes, seemingly without resting the whole foot on the ground. In these film extracts, this makes the young Elvis appear slightly unsure of himself as a dancer. Anderson devised an overall structure and created arm and upper-body movements to go with the legs, hence the piece's title. Some of these torso and arm movements came from Presley's own performance, while others suggest the extrovert performance of heavy metal and glam rock stars: pointing towards the audience or into the wings, strumming an air guitar and other larger-than-life rock star poses. Crucially, Anderson also gave the performers the direction never to acknowledge the beholder nor to exchange glances with one another, but to perform as if practising steps on their own in front of their bedroom mirror. While the 'dago' in *Façade* highlights a moment when men were no longer learning couple-based social dances from dance teachers, *Elvis Legs* refers to the solitary way that, from the late 1930s, people began to learn to dance to jazz and pop music (see McQuirter 2002).

In addition to this fragmentation between upper- and lower-body movement and a deliberate deconstruction of the kind of charismatic presence associated with lead singers at rock concerts, *Elvis Legs* has a minimalist choreographic structure. For the first two-thirds of the piece, all the movement is performed at the same pitch of energy, then for the final third stepped up to another higher energy level. There is thus no sense of development or climax. The dancers were spaced out along a deep diagonal line and remained dancing on the same spot throughout the piece. First, one after the other, each danced their own solo. Then these were repeated in sequence by the other two dancers while the initial soloist jumped from one rock star pose from their solo to another. All then turned to face backstage and the sequence with freezes was repeated again. Turning forwards once more towards the audience, they performed the freezes material once last time and then finally, as the music changed to a much faster tempo, danced all three solos right through in unison. This, and the moment at the start where each dancer introduces their moves, are the only times when the beholder gets to see the movement clearly. For most of the rest of the piece, it is only seen in fragments and in frozen moments. This abstracted fragmentation robs the movements of their narcissistic potential by encouraging the beholder to appreciate the dancers' attention to detail rather than focus on any overall characterisation.

Steve Blake and Drostan Madden's music samples guitar riffs and bass lines from rock standards so that it sounds deeply familiar, yet remains curiously nondescript, always seeming about to introduce a well-known tune that never emerges. For the Featherstonehaughs's performances this was played on guitars and drums while for the Laban version Blake arranged it for a brass 'oompah' band. The music set the mood and provided the dancers with the support they needed in terms of rhythm, and cues, never becoming prominent enough to be something they were dancing to. This choreographic and musical minimalism, together with the self-effacing performances and fragmented movement material, gave the traces of Presley's popular, iconic dancing an ironic humorous quality.

As I have noted, the dancers in *Elvis Legs* seemed both inside and outside what they were doing. While one recognised their ability to create Elvis-like moments, one also appreciated that on another level the dancers were equally interested in the ironic, detached quality of their performance. This quality was particularly strong in the women's versions. Whereas the men had worn glamorously flashy silver suits, the women wore tailored trousers and jackets. These did not make them look like Drag Kings but suggested, perhaps, the aristocratic suited lesbians painted by Romaine Brooks in Paris in the 1920s. The three women dancing in the Laban version were each very different kinds of dancers: Anna Pons Carrera is Spanish, Teresa Barker is British, and Maho Ihara is Japanese. Anderson frequently takes dancers with very different body types and artistic temperaments and rehearses them so that they perform very detailed movement material in perfect unison. This paradoxical unison was particularly evident in the women's version of *Elvis Legs* where the dancers seemed to bounce lightly but precisely through the work. The result was that while one enjoyed the women's beautiful performance of stereotypically masculine, Elvis-like pelvic movements and rock star arm gestures, one was equally aware that these were definitely not being performed by men but by women who were, in a strangely indefinable way, both feminine and unfeminine. One recognised both the energy and enthusiasm with which they took over *Elvis Legs* and made it their own, while on another level appreciated that they were relishing the sophistication and humour of what they were doing.

A preposterous comparison

I have examined a very recent piece which exists on DVD (and which I was able to watch in rehearsal) but which has not received much serious critical attention, with one from the past and for which there is a wealth of available information. On a methodological level, it might seem problematic to make a comparison between these two pieces when the kind of information I have been able to draw on in each case is qualitatively so different. The idea of comparing two works from different periods is one which is the starting point for most of my chapters in this book. This dialectic approach is not intended to function in a Hegelian way: it is not used as a way of deriving a new synthesis from contrasting theses nor of uncovering a transcendental aesthetic spirit progressing through history. Instead, this approach is, in part, inspired by the concept of preposterous history proposed by the Art Historian Mieke Bal. In her book *Quoting Caravaggio*, Bal discusses the way recent quotations from Baroque painting in the work of late twentieth-century visual arts can help us understand 'the cultural processes that integrate the past into the present' (1999: 3) and allow us to revision the Baroque in new ways. In a similar manner, I suggest, Anderson's *Elvis Legs* opens up possibilities for revising our understanding of the position that the tango occupied within *Facade* and its relevance for the present. This is therefore a preposterous comparison between a ballet by a canonical choreographer responsible for establishing the choreographic identity of the Royal Ballet, and a contemporary dance piece by an independent choreographer who for over 20 years has succeeded in maintaining her two companies outside the institutionalised dance world. It is thus a comparison that deliberately ignores a canonical account of (English) ballet history by blurring the boundaries between mainstream and fringe.

It would be a mistake to think that one recognises familiar liberal values within Anderson's *Elvis Legs*, and then criticise Ashton's *Façade* for reinforcing xenophobic and classist discourses through its performance during the 1930s. It is necessary, instead, to acknowledge the extent to which the dancers and choreographers in each historical example were working within particular social and political contexts. The women's version of *Elvis Legs* gave the dancers opportunities to perform in a strong and interesting way. Lopokova's role, in a very different social context, allowed her to imply that, as Judith Mackrell has shown, she was not as stupid as some members of the Bloomsbury Group considered her to be.

As I noted earlier, there are invariably instances where gaps and inconsistencies open up within dominant ideologies. *The Times'* critic, as I pointed out earlier, felt compelled to issue a warning about Ballet laughing in the wrong place. This indicates that in 1931, some aspects of *Façade* troubled social norms rather than confirming them. The critic didn't elucidate what he meant by 'the wrong place', thus appealing in a rhetorical way to common-sense notions of good taste so as to warn the Ballet not to infringe them. Humour, however, is too elusive to be susceptible to this kind of censorship. Because it has the potential to make people aware of shared social values, it can often reveal the way such values shift and change before such changes appear in other social registers. *Elvis Legs* and *Tango Pasodoble* demonstrate how adept Anderson and Ashton have been at using dance to convey these shifts and changes by drawing the beholder's attention to subtle nuances of performative interpretation. The pleasures that these two works offer to their beholders thus derive from the deft and witty way in which they open up a space in which it becomes possible to think identities differently.

Notes

1. For further information about The Camargo Society, see Kane and Pritchard 1994. An earlier version of my discussion of *Façade* appeared in Ramsay Burt 2002, and I am also grateful to Professor Gabrielle Klein for inviting me to present at the conference *Translation Dance* at the University of Hamburg, November 2007, where I was able to explore the humorous side of the Tango.
2. Keynes did not, however, take over the role of treasurer until after the first performance of *Façade*. See Kane and Pritchard 1994: 64. See also Kavanagh 1996: 114–15.

3. Lynn Garafola notes that while many members of the Bloomsbury Group had enjoyed performances by the Ballets Russes before the 1914–18 war, it was after the war that they became particularly enthusiastic about the company's work: 'What delighted Bloomsbury, above all, was the visual modernism of the troupe's newest productions. For the first time, Diaghilev used his proscenium to frame genuine works of art; he turned the stage into a magnificent art gallery serving the artists Bloomsbury had championed for a decade' (1989: 336).
4. Angela Kane and Jane Pritchard have tracked the details of this transaction in the minutes of meetings of Camargo Society committees. On 19 March 1931 Rambert offered to pay three-quarters of the design costs for the ballet. This was while the ballet was still being created. On 30 April, three days after its first performance, she paid £25 'for hire not ownership'. Then on 2 June, because performances of *Façade* at the Lyric Theatre were deemed to be a commercial undertaking, she was asked to pay a 15-shillings fee per performance. The society was evidently hard up, and not perhaps on particularly friendly terms with Rambert. It also suggests that while, as I shall show, most of the ballet critics either ignored or discounted *Façade*, Rambert recognised its value from the start. See Kane and Pritchard 1994: 63–4.
5. *The Times*, 27 April 1931, p. 10.
6. The *Daily Mail*, 27 April 1931, p. 9. The *Daily Sketch*, 27 April 1931, p. 3. The *Daily Telegraph*, 27 April 1931, p. 8. The *Manchester Guardian*, 27 April 1931, p. 13. No review of the ballet appeared in the *Daily Express*, the *Observer*, or *The Listener*.
7. A grant from the Arts and Humanities Research Board in 2003 enabled me to observe all the rehearsals for the Laban version of *Double Take*. The twentieth anniversary version in 2004 had two halves. In the first, male dancers performed Anderson's 1989 piece *Flesh and Blood* which was made for the Cholmondeleys and had been revived by them in 1997. This gender reversal matched the women's performance of eight pieces initially created for the Featherstonehaughs (this was two more pieces than had been performed at Laban).

5

Napoli and *Palermo, Palermo*: Cosmopolitanism and Energetic Excess

Ramsay Burt

This chapter compares two dance pieces which performatively recreated memories of visiting a Southern Italian city: the 1842 ballet *Napoli* by the Danish choreographer Auguste Bournonville (1805–1879) and the 1989 dance theatre piece *Palermo, Palermo* by the German choreographer Pina Bausch (born 1940). These are two very different works, each created in different contexts. In the 1840s both Denmark and the Kingdom of Naples were absolute monarchies and still largely untouched by the Industrial Revolution that was already bringing about rapid social and economic changes in other European countries. In the Ruhr district where Bausch's company are based, German economics and demographics, during the last 40 years, have necessitated the temporary economic migration of foreign 'guest workers' from Mediterranean countries. Putting these two works together brings out significant correlations between them. In each case, the idea of the paradisal otherness of Mediterranean culture exemplified by a hot southern city was created through theatrical staging and movement material that embodied a hot-blooded utopian dynamism and energetic excess. I shall argue that, in each case, this excess compensated for a sense of loss arising from social and ideological rupture – in Bournonville's case in relation to Romanticism, and in Bausch's case, modernism.

This chapter uses the comparison of these two dance pieces from contrasting historical periods to interrogate the different ideologies of cosmopolitanism that each performed. The value of these two dance pieces, I suggest, lies in the warmly positive way they each embrace foreignness. At a time when public concern in Western Europe about immigration is inspiring fears about the loss of national identity, these two pieces can help us revisit eighteenth-century discussions about cosmopolitanism. They remind us about the persistence of some liberal eighteenth-century ideals, and help us reflect on the significance these had in the 1840s and the meaning they have for us at the start of the twenty-first century. While the word 'cosmopolitan' generally refers to the benefits of knowing about countries other than one's own, in the late eighteenth century, philosophers in France and Germany used it in connection with the problematic of human rights. Whereas human rights ought to be universal, individuals only actually enjoyed them as citizens of modern nation-states, thus raising the problem of the rights of foreigners. In his 1795 essay *Perpetual Peace*, Immanuel Kant argued that foreigners had a cosmopolitan right, as he called it, to hospitality: 'by virtue of their common ownership of the earth's surface; for since the earth is a globe, they cannot scatter themselves infinitely but must finally tolerate living in close proximity' (Kant 1983: 118). One example Kant gave of intolerance towards difference was the behaviour of colonial powers who regarded Africa, India, and America 'as land belonging to no-one because their [indigenous] inhabitants counted for nothing' (ibid.: 119). By the late twentieth century, this intolerance to difference had moved closer to home. The art historian John Berger, in his 1975 book *A Seventh Man*, protested against the institutionalised dehumanisation of migrant guest workers in northern European countries. In the 1990s and 2000s, Derrida (2002), Kristeva (1991, 1993) and others revisited the eighteenth-century discourse on cosmopolitanism, including Kant's essay, in order to critique the treatment of stateless refugees and economic migrants.

Kant's argument about tolerance is one that negotiates between rational and subjective judgment. It is a logical proposition that increasing population will lead to people living in ever closer proximity with one another, but what 'too close' might mean is essentially a subjective judgment. In Kant's philosophy, the First Critique, which deals with epistemology, and the Second, concerning ethics, are brought together in the discussions of aesthetic judgment in the Third. While this is generally accepted, the role which ethics must therefore play within aesthetic judgment is rarely noted. Although the Third Critique is concerned with natural beauty rather than with art as such, for a younger generation of Romantics, art became the privileged site where the crisis of the modern world could be resolved. As Simon Critchley points out: 'The problem to which Romanticism attempts to find a solution is that of how to reconcile the values of the Enlightenment – secularisation, humanism, the libertarian and egalitarian values of republicanism, the primacy of reason and the ubiquity of science – with the disenchantment of the world that those values seem to bring about' (2004: 100). Put in this way, Romantic disenchantment is a precursor for the affective consequences of the rupture of modernism. Walter Benjamin evoked the latter in an often quoted statement about the emotional and psychological impact of the First World War: 'A generation that had gone to school on a horse-drawn street car now stood under the open sky in a countryside in which nothing remained unchanged but the clouds, and beneath these clouds, in a field of force of destructive torrents and explosions, was the tiny, fragile human body' (1973: 84). Awareness of the vulnerability of the human body and of one's responsibilities towards the vulnerable are ethical matters. The relation that Kant created between epistemology and ethics is also therefore active within Benjamin's philosophy.

For Bournonville's audience, as for Bausch's, art became an area in which to imagine what the world might be like if its existential crises could be resolved. My intention in bringing together *Napoli* and *Palermo, Palermo* is not, however, to read them as symptoms of social and political problems, nor to use these to analyse the social moments that produced them. My objective is to analyse the works themselves, but my starting point is that one needs to acknowledge how a particular society defines the function of art in order to be able to gain a deeper understanding of the value and significance of an art work. This, I contend, is partly revealed within the ethical stance which the work performative constructs. In a critique of the deleterious effects of globalisation at the beginning of the twenty-first century, Zygmunt Bauman (2003) has observed that the stateless migrant and the globetrotting multinational businessman are both somewhat homeless. This is because they are in the position of being in, but not of the space they physically occupy, whether it is a refugee camp or a Hilton Hotel. At stake for Bauman are the conditions that permit or hinder individuals from acknowledging the existence of differences. My aim in comparing a nineteenth-century ballet with a twentieth-century dance piece is to identify how each responds to the ethical implications of the collapse of the romantic distance between the south of Italy and northern Europe.

Bournonville was employed by the king of Denmark from 1830 until his retirement in 1877, as choreographer and theatre director at the King's Theatre in Copenhagen where *Napoli* was first produced. In his memoirs, *My Theatre Life* (1979), Bournonville writes that in 1840, following the unfavourable reception of his ballet *The Toreador*, he considered pursuing his career outside Denmark. He therefore took leave of absence to travel through France and Italy. His six-week stay in Naples inspired the ballet, *Napoli*, which some consider his finest achievement. Bournonville's contemporary letters and later memoirs reveal how the Mediterranean appeared to his northern eyes. Kant's essay *Perpetual Peace* was written at a time when artists and writers from northern Europe, including Goethe, Byron and Turner, were travelling through Italy to stimulate their creative vision through romantic and sublime experiences. It was this which Bournonville, like a number of Danish artists, sought through a trip to Italy; but 1842 saw not only the first performance of *Napoli* but also the publication in Leipzig of the first of Karl Baedeker's famous guides. As Roger Cardinal has pointed out: 'Now, in the space of a few years, the elective sites of an intellectual elite were being devalued, de-sacralised, recycled as so many perfunctory stop-overs on a pre-established circuit' (1997: 148). It is clear from his letters home to his wife and his later memoirs that this circuit is to some extent already in existence. The blueness of the Blue Grotto in Capri, Bournonville wrote, was 'so striking to us [from Denmark], and far surpasses the colour of our Sound and lakes, even when our sky is completely Italian' (1979: 94). In Rome and Naples, he wrote, 'art, antiquity, and the Middle Ages confronted me at every step but [...] the folk life and beautiful scenery in particular captivated my imagination' (ibid.: 531). The fact that Bournonville seems to have had a stronger response to folk life than to classical culture suggests a Romantic sensibility. Folk life, of course, included folk song and dancing.

Napoli: eller, Fiskeren og hans Brud (Napoli, or the Fisherman and his Bride) was a narrative ballet about the obstacles and setbacks faced by a poor but worthy young fisherman Gennaro (initially danced by Bournonville himself) and Teresina the girl he loves. In his memoirs, Bournonville noted that the little plot was but a single strand in the braided garland of the ballet in which he wished to present Naples just as it appeared to him (1979: 95). The first act showed a colourful and busy street scene at the waterside fish market on the Quay of Santa Lucia in Naples and included a spectacular storm scene. The Blue Grotto was shown in the second act where Teresina, having fallen out of Gennaro's boat in the storm, found herself in the clutches of the evil sea spirit Golfo. Gennaro of course rescued Teresina. The third act showed the Whitsun procession returning to Naples from Monte Virgines. Here Gennaro and Teresina were married by a friar, after which the young men and women of Naples celebrated by performing a long, complicated suite of dances inspired by the folk dances Bournonville had observed in Italy.

I have based my analysis of Napoli from viewings of a 1986 video recording of a performance on stage by the Royal Danish Ballet. This must not, of course, be taken as a wholly faithful or entirely authentic version of the ballet as it was performed in Bournonville's day. (Nor should the experience of watching video be conflated with that of watching live performance.) As Jeannette Anderson has noted, the ballet was restaged in 1901 by Hans Beck, and later in the twentieth century by Harold Lander, Hans Brenaa and Kirsten Ralov. One can, however, cross reference some aspects of the ballet as it exists today with Bournonville's own description of it in his memoirs. While in Naples, Bournonville had stayed in a room overlooking the Quay of Santa Lucia, and many of the details of the ballet's first act correspond to the incidents of daily life beneath his window as he described them, at the time, in letters to his wife (McAndrew 2002).

The first act of *Napoli* presented a foreign and picturesque scene, but one that was also busy, modern and metropolitan. The actions of the principal characters in the story were sometimes almost in danger of being lost in the details of the waterfront with its macaroni cooks, fruit sellers, housewives, and fishermen, and in the excitement caused by the arrival of a travelling puppet show. Within the conventions of the ballet stage, this was perhaps a *flâneur*'s view of city life as a series of alien fragments. Walter Benjamin seems to suggests that a Baudelairean *flâneur*'s experience of the modern city was almost that of a foreigner. Bournonville was, of course, a foreigner in Naples. The backdrop for this scene was the bay of Naples with, in the distance, the smoking volcano, Mount Vesuvius, a persistent reminder that nature was more volatile and impulsive here than in northern Europe.

The first act's fragmented view of city life was further disrupted by the storm with its carefully choreographed chaos, as dancers careened around the stage as if blown by the wind. The narrative function of the storm was to separate Teresina from Gennaro, while on another level it signified a disruption of social harmony that would be healed, in the third act, by the marriage and its accompanying celebrations. Thus the sublime, natural cataclysm of the Mediterranean storm was recuperated by the Mediterranean intensity of the ballet's final divertissements. These, however, were located just outside the city, with the central arch of a stone bridge dominating the backdrop, so that the *mise en scène* was neither countryside nor townscape but somewhere undecidably in between the two. Similarly, the dancing was neither folk dance nor ballet dancing but a hybrid that suggested both the 'natural' energy of the one and the refined shapes and proportions of the other.

The inspiration for the divertissements can be found in one of Bournonville's letters to his wife. This describes an afternoon when, having gone with a party to view Roman ruins outside Naples, he bought some sea shells and mosaic fragments from a village girl, Rosa. He writes that she 'surpassed the other [girls] in liveliness and was very pretty into the bargain'. He encouraged her to dance for them which she did together with some other young people:

> One banged the tambourine, another sang: and now Rosa and another girl began the tarantella. It was no masterpiece; it was performed in bare feet, and the steps were rather monotonous. But the location, the air, the delicious scent of flowers from the surrounding gardens made it extremely poetic. It was a changing dance; a young chap called Rafaelle took the place of the second girl and performed many comical movements. Encouraged partly by my travelling companions [who knew Bournonville's profession] and partly by my own irresistible urge, I tossed aside coat and neckerchief and joined in the dance. Rafaelle stepped aside; and I now performed all the figures I could remember. Everyone joined in the singing, the tambourine's tempo was doubled, Rosa laughed and shrieked. I danced until I could no longer stand it, and we ended the dance to universal cries of 'Bravo!' which rang through the lofty vault of the ancient temple.
>
> (Bournonville in McAndrew 2002: 345–6)

Just as Bournonville couldn't resist joining in, so in the last act of *Napoli*, Gennaro leapt into the dance and took over from another man who was dancing with Teresina. (Teresina was perhaps getting her own back as, in the first act, Gennaro had teasingly danced with another woman.) In Italy, joining in was probably the best way for Bournonville to check the finer details of the dance, but it also demonstrates how much European folk dance and ballet had in common at the time. Theirs was, nevertheless, a hierarchical relationship, ballet sharing aesthetic values and a common gestural language with the academic traditions of painting and sculpture. It was this superior value that Bournonville must have assumed he was adding when he turned their 'monotonous' Tarantella into the divertissement at the end of *Napoli*.

While the Danish ballet dancers had a whole orchestra to accompany them, Rosa and her companions danced to a solitary tambourine and a human voice. Bournonville did not appreciate southern Italian folk songs, complaining in one letter about a female singer who screeched and wailed 'so that every vein in her forehead, neck and cheek became distended' (McAndrew 2002: 371). The Tarantella was a Sicilian dance, and Sicily was successively settled by Phoenicians and Carthaginians as well as Greeks and Romans. Nearly two centuries of Islamic rule had a lasting, beneficial effect on medieval Sicilian culture. Bournonville was an accomplished pianist but his musical education would not have included North African or Islamic music, let alone prepared him to appreciate any remnants of these that had been assimilated within southern Italian folk music. Were any residual traces of African and Islamic movement forms present in Rosa's performance of the Tarantella, and could Bournonville, despite his dislike of the music, have responded to them? Might he even have conveyed their affective quality to his dancers back in Copenhagen while creating the movement material for the last act of the ballet? While these questions cannot be answered, they indicate that it was through travel that nineteenth-century European citizens found themselves in situations in which they were confronted with the ethical problem of acknowledging difference. Bournonville's ballet suggests they sometimes responded in non-discriminatory ways.

I have stressed the closed, homogenous structure of nineteenth-century European dance culture because its breakdown in the twentieth century forms the context for Pina Bausch's Tanztheater. *Palermo, Palermo* is a more difficult piece to describe than *Napoli*. It began with a stunning theatrical moment when the curtains rose to reveal that the entire proscenium arch was filled with a five-ton concrete wall which then collapsed backwards on to the stage with a devastating crash. The rupture this brought about produced a theatrical environment that severely disrupted any possibility of performing 'normally'. For the next three hours, groups of dancers picked their way carefully through the rubble to perform a succession of scenes that are by turns humorous, painful or horrifying (see Figure 7).

Whereas Bournonville created *Napoli* with Danish dancers for a Danish audience, *Palermo, Palermo* was made by a company of dancers from many different countries and was co-commissioned by the Teatro Biondo in Palermo and Andreas Neumann International. Although, superficially, Bausch's work appears to have more in common with theatre than with dance, it depends upon the dancers' sensitivity to non-verbal aspects of bodily expression and to rhythm and pace. To research the piece, Bausch spent some time in Palermo with her company, Tanztheater Wuppertal. If Bournonville's Naples corresponded in some ways with the kind of tourist experience codified in Baedeker's guides, Bausch and her company's visit to Palermo belongs to a kind of modern tourism that Siegfried Kracauer sardonically characterised in a 1927 essay. For the modern traveller, he complained: 'one hotel is like

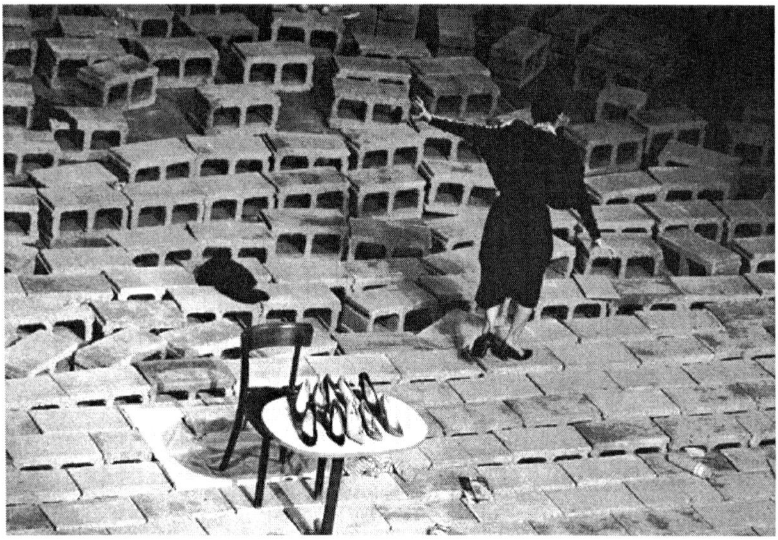

Figure 7 *Palermo, Palermo* Choreography: Pina Bausch, Tanztheater Wuppertal, Dancer: Nazareth Panadero
Photo: Jochen Viehoff.

the next and the nature in the background is familiar to readers of illustrated magazines [...] As a result of the comforts of civilisation, only a minute part of the globe's surface remains terra incognita today; people feel just as much at home in their homes as they do elsewhere – or they do not feel at home anywhere at all' (1995: 66). This feeling of homelessness is a consequence of Romantic disenchantment and modernist rupture. Bausch's work in many ways both responds to, and attempts to remedy such feelings of homelessness and makes us aware of ethics.

On returning to Wuppertal, Bausch worked with the company to make the piece, starting rehearsals on 13 April and finishing on 1 June of 1989. Bausch posed questions to her dancers, who put together short improvised sketches in answer to them. The piece was then assembled by selecting, editing, combining and structuring this material. The questions and themes which Bausch used during rehearsals were printed in the original programme along with photographs of street scenes in Palermo taken by company members. As Bausch told Claudia Provvedini, 'I just presented colours, sounds, figures, and emotions on the stage: perhaps there is violence, but there is also strength. And there is love at the core' (1990: 3). Viewing the Italian premiere, the Italian dance writer Leonetta Bentivoglio had no difficulty recognising the Sicilian city in Bausch's piece: 'Palermo, dreamed by Bausch, is a city that is both prostituted and holy, baroque and extremely violent, in harmony with history and infested with decay, seductive in its contradictions, which burn with the same intensity as the row of lighted candles, as in a church, which are supported by the arm of the man who plays a saxophone' (Bentivoglio 1990: 20). In the second half of *Palermo, Palermo*, a dancer, initially Francis Viet, lit a number of candle stubs that he attached to his bare, right forearm with melted wax. A female dancer, Quincella Swyningan, then asked him to play something for her, and he exited and came back with a saxophone. Despite his discomfort from dripping hot wax, Francis played her the mellow jazz standard 'Stormy Weather'. Meanwhile, behind him, another female dancer with a stocking mask over her face waved a revolver around in a menacing way. The row of burning candles, as Bentivoglio observes, evokes the interior of a Catholic church. The collision of fragmented images here might be interpreted as specifically Italian (baroque cathedrals, mafia) but could equally well be seen as international or even cosmopolitan.

Palermo, Palermo was one of Bausch's messiest pieces. Not only was the stage strewn with broken rubble, but dancers, throughout the evening, kept on dumping more and more mess on to it. Overripe tomatoes, aimed at Julie Shanahan, landed on it (see Figure 11 in Chapter 7). Fruit thrown at corrugated-iron sheets leaning against the back wall bounced off on to it. Handfuls of fresh meat and raw sausages are dropped on to it by Jean Laurent Sasportes who retrieves them, bit by bit, from inside his clothes and down his trousers, as he is fiercely kicked in the bottom by a furious Beatrice Libonati. Red soil released from the fly tower, drifted down in a dusty cloud on to it. Food and bottled water was regularly spilt on to it. In a stately line, dancers process to the front of the stage to the sound of a Sicilian funeral march, letting clothes and fresh vegetables fall out of their arms on to the already littered stage with measured gravity. In the final few minutes, flowering cherry trees were lowered upside down on ropes out of the fly tower. The piece's spectacularly excessive waste robbed items of their value and disrupted the psychic or monetary economies in which they had previously circulated.

There are a number of scenes where female dancers take on a dominant, controlling role, using male dancers to assist them in activities that are often extreme, explosive or violating. I have already mentioned a moment at the start of the evening when Julie Shanahan calls on two men to pelt her with tomatoes. Valerie Briginshaw discusses this in more detail later in the book. Later, Beatrice Libonati comes on stage with three suited men as supporters, to express her contempt for another man. Jamming a litre bottle of mineral water, that one of her supporters has brought for her, between her thighs, she removes its cap and stares at the object of her contempt while its contents gush out on to the floor between them. With a wiggle of her hips, she deftly empties out the last drops and then turns haughtily and sweeps off-stage with her entourage at her heels. In a similar scene, Mariko Aoyama has an unexplained tantrum with Dominic Mercy and calls two men on from the wings. One gives her a wine glass and the other fills it with water before grasping her from behind and spinning her quickly upside down through 360 degrees so that the water in her glass is flung in Mercy's direction. It is as if these three women, like the soldiers in the trenches that Walter Benjamin described, have been reduced to no more than vulnerable, sentient beings within a field of destructive social and political relations. In contrast to their explosive outbursts, the supporting men do what is required of them, but remain relatively neutral, like witnesses. Recognising that these violent scenarios are unavoidable, they are nevertheless concerned about their consequences. Out of the conflagration of these volatile displays of female will come the ashes of male compassion and even perhaps tenderness.

Palermo, Palermo was first performed on 17 December 1989 when the demolition of the Berlin Wall, which began on 9 November, was still fresh in people's minds. This led some critics at the time to connect the opening of the piece with the momentous events of German reunification (e.g., Jeitschko 1989; Kayser 1990; Koegler 1990), an interpretation that Bausch's long-term collaborator, the scenographer Peter Pabst, strongly denies. The rehearsals and plans for the decor, he points out, had been set in train long before November. The idea, however unauthorised, that the opening sequence relates to the fall of the Berlin Wall has nevertheless stuck in some people's minds. The danger here, however, is of reducing this to only one level of meaning and missing other overlapping signifying layers which the falling wall also evoked. As Bausch herself observed: 'There are many walls to be brought down. I think that we are all in some sense held back, constrained. And at the same time we all want to be loved or liked by others. I have tried to express this in my dance, but perhaps you might say also in other ways' (Provvedini 1990: 3).

Pabst, who explains the genesis of the idea for the falling wall in Chapter 8, used real bricks because he wanted a really solid bang – an earthquake as he has called it. His reference to earthquakes exemplifies differences between the Mediterranean and northern Europe. A seismic fault line runs through the entire Italian peninsula from Sicily, along the Apennines to the Eastern Alps, causing frequent earthquake tremors as the continental plates of Africa and Europe shift. Familiarity with travel magazines and television programmes means that *Napoli*'s decor and its choreographed storm may seem no more than historical curiosities to modern viewers. The sensational violence of *Palermo, Palermo*'s 'earthquake' disrupted familiar expectations of theatre dance, putting audiences back in touch with the strange and sublime experiences that Bausch and Bournonville had both found in the Mediterranean.

The rupture of modernity which these natural disasters suggest is one which challenges and tests the limits of European cosmopolitanism. While *Palermo, Palermo* may have had the Italian qualities which Lionetta Bentivoglio described, these were part of a multi-layered, fragmented assemblage that could be described as multicultural. This is particularly evident in Bausch's eclectic choice of music most of which would not have been appreciated by educated Europeans in Bournonville's day. This includes American jazz numbers like 'Stormy Weather', music by Tchaikovsky and Paganini, drumming from Japan, thumb piano music from central Africa, a Jew's harp and a recording of precisely the kind of Tarantella song that Bournonville disliked. Bausch used the Tarantella as the musical accompaniment for a busy scene at the end of the first half. While stage hands and dancers tidied up the rubble and rubbish, other dancers took it in turns to come forward and face the audience to perform almost violently energetic sequences of arm and hand gestures that seemed propelled by the song's powerful rhythm. Compared with northern Europeans, Italians appear to use a particularly rich and emphatic vocabulary of gesticulations to emphasise their conversations. After a while, little duets occurred as dancers faced one another in what began to appear like non-verbal arguments, their arms held high, flinging gestures with arrogant intensity at one another with, perhaps, the stinging menace of a tarantula. The movement material in both Bausch's and Bournonville's tarantellas therefore exemplified the embodiment of a hot-blooded utopian dynamism and energetic excess that tourism has established as characteristic of the Mediterranean region.

Siegfried Kracauer argued that modern travel had undermined the difference between home and abroad so that, repeating his observation, 'people feel just as much at home in their homes as they do elsewhere – or they do not feel at home anywhere at all' (1995: 66). Naples in 1840 was just far enough from Copenhagen to seem strange, but not so far as to become exotic. Bournonville felt he had enough in common with the young villagers to join in their dance. For social and historical reasons, had he been travelling in Morocco or Egypt, he would almost certainly not have been able to do so. Where Romantic ballets were set in non-European countries, difference was always objectified as exotically Other. Bournonville, however, created a world for his ballet in which the difference between Denmark and southern Italy was appreciated in an ethical, non-discriminatory way that accorded with Kant's concept of cosmopolitan hospitality. In Bausch's piece, the disorienting, fragmented mixture of references to different cultural traditions was one that Bausch and her dancers were as likely to encounter in Germany as abroad. Modernity and globalisation had collapsed a feeling of distance that existed in Bournonville's day. *Palermo, Palermo* performatively created an experience of not feeling at home anywhere at all because the ethical challenge of cosmopolitan tolerance was now posed everywhere that people from different cultures lived in close proximity to one another, and that included both Palermo and Wuppertal. Yet if this piece, as Leonetta Bentovoglio suggested, was Bausch's dream of the city of Palermo, it was a vibrant yet violent dream of spectacular excess. Rather than fixing and exoticising Sicilian otherness, this excess evoked historically and culturally changing identities in a performative superfluity that took no single form and told no single narrative. By doing so, such openness offered the hope that a solution to the ethical problem of facing up to cultural differences might be imagined through dance performance. To conclude, *Napoli* staged an imaginary Mediterranean in which to present homeliness at a distance by affirming cosmopolitan tolerance. *Palermo, Palermo,* at a time when such distances seemed to be collapsing, performatively reconstructed fragmented memories of the Mediterranean in order to generate experiences in which to provoke individuals to remember their own potential for openness and ethical responsiveness by having it reflected back to them in sensuous images of violent intensity.

Section 3
Uncontrollable Intensities

Introduction

At first sight it might seem odd to bring together in the same section the work of, on the one hand, Lea Anderson and, on the other, Pina Bausch. However, both artists exploit all the physical resources of the modern stage including set, music, costumes and lighting, to create the magic of theatrical artifice, at a time when this seems to be out of fashion within contemporary choreography. This helps to create the distinctive moods that colour the performances in these choreographers' works. What these two chapters both emphasise is that in these performances, each choreographer foregrounds the materiality of the dancing bodies. The result of this can sometimes seem uncomfortable. Our chapter presents an alternative reading of Anderson's *Yippeee!!!* to the critics' notices on its premiere. It argues that its scandalous mining of Busby Berkeley's dance routines produced radical and subversive intensities. Valerie's reading of three female solos from Bausch's repertoire, through exploring the open corporeality of bodies that cannot be otherwise, discusses the ways in which the dancers are irreparably exposed. Here lies the difference between the scurrilous humour of Anderson's work and the intense seriousness of Bausch's. What makes these artists not only important but also exciting as choreographers is the way that they use choreographic innovation to resist reinforcing cultural and ideological norms. Their resistance is bound up with the ways in which Anderson's work explores failed copies, and Bausch's work presents failed bodies. In the fissures that these strategies open up, uncontrollable outpourings of expressive energy and intensities are revealed. What is uncontrollable in Anderson's work is the subversive humour that drives her choreography, whereas, in the three Bausch solos, it is the intensity of the dancers' performances that cannot be contained. Despite the very British humour underlying the former and the deeply serious tone of Bausch's solos, what both chapters reveal is that these artists are keeping open the potentialities that allow us to imagine alternative ways of existing.

6
'Don't mention Busby Berkeley': A Reassessment of Lea Anderson's *Yippeee!!!*

Valerie A. Briginshaw and Ramsay Burt

Introduction

Yippeee!!!, a 90-minute dance created by Lea Anderson in 2006, is a piece that exemplifies many of the themes that we have discussed in other chapters, particularly concerning multiplicity, perverse hybridities and ironic humour. Where we discuss these issues elsewhere, it is usually in the context of conceptually based anti-theatrical European work from the last ten years. What's distinctive about *Yippeee!!!* is that it is actually highly theatrical, drawing on popular culture with a wealth of spectacular and sophisticated costumes by Simon Vincenzi and a live jazz rock score by Steve Blake. One of Anderson's starting points for her choreography was dance sequences in films by Busby Berkeley from the early 1930s, her title *Yippeee!!!* echoing that of the 1930 film, *Whoopee!*. Berkeley's frivolous title seems ironic when one realises it was made at the height of the depression. As Anderson told Martin Hargreaves, the same applies today: 'the show is about now and it's quite cynical – it's also quite silly – but there is definitely a question of what is there to say "Yippeee!" about now?' (2007: 26). On the one hand it is quite accessible through its silliness and its references to popular culture, but underlying it as Anderson implies, are much more serious concerns.

Anderson has been a feature of the English dance scene for over 25 years. We have both been writing about her work throughout this period. However, for readers in continental Europe and the United States who may not be so familiar with her work, we offer the following brief introduction. The words 'streetwise' and 'urban' regularly appeared in reviews of Anderson's work particularly in the early years. Whereas contemporary dance work is often serious and a little difficult and sometimes obscure, Anderson, as she told Barbara Newman in 1987, wanted to 'appeal to a much broader cross-section, people who are theatre goers or music goers or even just people who go out, night club people' (1987: 130). Much of her work takes images or short extracts from sources such as Hollywood movies, rock concerts or advertisements and recreates them within highly stylised choreographed sequences.

In 1984 she formed her first company, the Cholmondeleys who were all women, and in 1988 she formed the Featherstonehaughs, an all male company who were, in effect, a boy band before that phenomenon emerged. Her work with each of these single-sex companies was informed by her astute critique of the norms of gender and sexuality circulating within contemporary culture (both companies combine to perform *Yippeee!!!*). Missing from almost all of the critical reception of Anderson's work has been any recognition of her preoccupation with the formal qualities of abstract movement and symmetry and with the expressive qualities of contemporary dance movement vocabulary. Because her work is assumed to be accessible, critics seem not to notice that there is much more going on in it. Anderson herself is well aware of this. Furthermore, as she told Ramsay in 1991: 'just get everyone all buddy buddy, get everyone all nice and relaxed, you know – the Featherstonehaughs, all nice and accessible – and then just start to make them uncomfortable' (in Burt 1995: 169). Formal and expressive qualities in Anderson's work that trouble audiences and the underlying issues they point towards have always interested us, and these are particularly evident within *Yippeee!!!*.

This inability to get beyond the superficial connotations of popular subject matter was particularly noticeable in the reviews of *Yippeee!!!*. While the show received some favourable reviews, the critics in the major newspapers seemed to have approached *Yippeee!!!* with similar preconceptions of what a piece based on Busby Berkeley numbers would look like. As Judith Mackrell wrote in the *Guardian* '*Yippeee!!!* came advertised as a homage to Busby Berkeley, it seemed we were in for a visual treat' (2006: 36). The reviewers seemed to concur that the piece had too much repetition and was too long, including phrases such as: 'dismal and heavy handed'; 'sinister and dull'; 'without momentum and flourish' (Anderson 2006; Craine 2006: 17). Zoë Anderson concluded that the choreographer then 'misses the big weird scale of those old films, their surreal exuberance' (2006). She appeared to want simple, straightforward transcendence rather than the darker images Lea Anderson had mined from the production of Berkeley's films. As

Lea told Hargreaves: 'Berkeley went through hundreds of dancers and Simon Vincenzi found photos of them, between shoots, collapsed and crying, queuing to see the doctor' (2007: 26). One of Vincenzi's more grotesque costumes trussed the dancers' chests and arms stretched above their heads in bandages, revealing at the top red-gloved hands suggesting bleeding fingers. We also see the dancers staggering and exhausted at the end of a sequence based on the number 'Young and Beautiful' (from *42nd Street*); they are dizzy because they have mimetically suggested they were dancing on concentric turntables with lots of spinning movements. Zoë Anderson complained that there was a lot of milling around in the dance sequences 'which looked half-finished' (2006). She clearly missed the point that *Yippeee!!!* was revealing what also happened immediately the cameras stopped rolling. In her view 'it builds energy but not momentum', whereas we suggest that it builds a tellingly uncomfortable intensity. It's the ironic and humorous ways in which *Yippeee!!!* generates these intensities, and, in doing so, explores multiplicities and suggests hybridities and mutations, that we address in the rest of this chapter.

On-stage/off-stage

The way the stage is set in *Yippeee!!!* exemplifies one of Anderson's main strategies for developing her revisionist take on Berkeley's work. The stage space is marked as on-stage – where spectacular, set piece, group numbers and concentrated, intricate duets are performed – and off-stage – the clearly visible wings where rails of costumes and accessories stand ready for the show's many costume changes. Demarcating the boundaries between the two are eight-foot-high towers of stage lights on casters which the dancers arrange and rearrange in different formations: two facing lines, a 'V' shape, an oblique diagonal or a receding semi-circle. As in most of Anderson's work over the last decade, what happens off-stage is as carefully choreographed and rehearsed as the main numbers. Dancers quench their thirst by swigging from champagne bottles. One smokes a cigarette while absorbed in watching another mark through a few steps. There is nothing sentimental about this off-stage activity. Dancers do not reveal their 'true' characters or personalities but remain as impersonal and anonymous as they do when dancing the stylised movements of the main numbers. Particular attention is given to transitions and rearrangements of the lighting, which the dancers execute in an economical and efficient way. In *Yippeee!!!* the relationship between on- and off-stage becomes one between different time-spaces when we see what happens when the 'cameras' stop. This is an instance where time comes before space because the dancers appear to have shifted to a different kind of time-space as a result of a temporal displacement rather than a spatial one. This relationship between stage and wings, when the dancers drop out of a rarefied performance mode into another more mundane one, is crucial to the way *Yippeee!!!* produces its particular theatrical qualities.

While the virtual cameras 'are rolling', the dancers try to attain the ideal geometric precision that is demanded of them, but, during the breaks between, reveal the underlying cost to them of this attainment. These transitions are key factors contributing to *Yippeee!!!*'s intensity. The more clumsy and casual the actions in the wings appear, the purer and more ideal the dancing on stage becomes, but only because of the relationship of dependence between the two. It is these clumsy, material bodies that create the illusion of disembodied ideality. This is why the transitions are so important. When the dancers step out of formation and relax their posture, the beholder is suddenly reminded of the dancers' beautiful shapes and of the elegance of the composition as a whole precisely because it is disintegrating. *Yippeee!!!* insists on the interrelationship between these two distinct but interdependent levels. This is not a dialectical synthesis which supersedes the merely human to attain a transcendent ideal. (This is what some of the critics had anticipated.) It is a challenging reconciliation between two seemingly incongruent levels that results in a new kind of harmony. Anderson discovers different kinds of aesthetic qualities from those normally associated with Berkeley's films, and suggests different ways of reading them.

Humour

While aspects of *Yippeee!!!* suggest a darker, perverse side to Berkeley, it nevertheless also reveals the silliness of much of his choreography. This is also a consequence of the way Anderson plays with different modes of performance staged in different kinds of conceptual spaces. Much of the piece's ironic humour involves such plays. Irony involves stating one thing in a rhetorical way that implies the opposite, often with humorous effect. Dancers frequently convey an awareness that although what they seem to be doing may be entirely innocent, it really means something quite different. These moments often involve smiling towards the audience, for example while incessantly repeating the same 'step ball change' sequence, or wiggling their hips in unison in a coyly sexual way. These smiles are often a little too shiny and toothy because the dancers are actually wearing gum shields to which chromium teeth have been attached (see Figure 8).

While some of the humour depends on the dancers' knowingly ironic performances, it also comes from *Yippeee!!!*'s hyperbolic exaggerations, particularly within the costumes. While these suggest Hollywood glamour, they seem also to hint at the perverse. Skin-tight white body suits are a little too transparent. Not only can the audience see the dancers' prosthetic pubic hair underneath but this is over-abundant and even seems to move to different parts of the body between numbers. One especially crazy number has the dancers dressed as sperms, with shiny white protuberances across their stomachs and extended bottoms, to which long thin tails are attached which the dancers spin and twirl with wicked glee. In 'Pearl Poseurs', Maho Ihara glides elegantly forward wearing a beautiful, white, art deco costume with symmetrical swags of pearls hanging from just above each elbow to a choker round her neck. It is slightly disturbing however that her two black-clad attendants are wearing neat little black gas masks. On the stage's periphery at each side, two more dancers briefly appear in white tights and gas masks from which

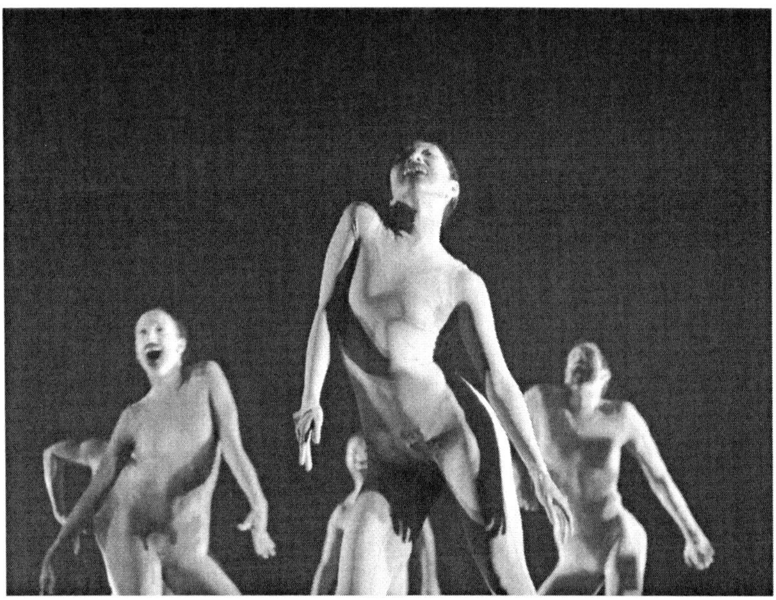

Figure 8 Lea Anderson's *Yippeee!!!*: smiles that are too shiny and toothy
Photo: Pau Ros.

sprout fetishistic latex tubes that, like Ihara's costume, hang in swags to just above their elbows.

Multiples and failed copies

One of the reasons Anderson chose Busby Berkeley films as a starting point for *Yippeee!!!* was the opportunity it gave her to focus on multiples, patterning and copying. She indicates that she was interested in 'thinking about multiples over time, and about how when I sample something from a previous time period it will alter it [...] Each dancer will unwittingly change the material which can then spread – like in photocopying if there is a fold in the paper it becomes part of the image' (in Hargreaves 2007: 26). Anderson seems to be interested in the distinctly human factor which brings failure or flaws into the copies, like the fold reproduced in the photocopy of the photocopy. Whereas in the original Berkeley numbers, performers strive to perfectly fit into a pattern and imitate each other like automatons, hiding their human potential for failure or imperfection, Anderson transforms this by enjoying revealing our irreducible differences and the potential for productive repetition.

What we are calling multiples are productive repetitions that fail to reproduce the same. As failed copies they open up the potential for difference, for seeing things differently, for alternatives to the norm. For example, in the eponymous ensemble number from Berkeley's *42nd Street* (1933), the *Yippeee!!!* performers are costumed in shiny silk, puff-sleeved, knee-length smocks which only have fronts. The costumes present a front, but when the dancers turn around we see what is behind it; their almost naked backs in their body stockings. These half-frocks also point to the on-stage/off-stage doubleness, that the piece continually plays with. As is often the case in Anderson's work, these feminine, little girls' outfits, reminiscent of the Vivian girls' frocks in Anderson's earlier piece, *3* (2001), are worn by all the cast, male and female. In their feminisation of the male dancers, they fail to reproduce the same. Through revealing difference, they critique the hetero-normativity of Berkeley's original. The performers' various shapes and sizes are accentuated by the choreography's grid-like formations, where lines of performers graduate from the smallest at the front to the tallest at the back. By laying bare the differences of gender and size, these multiple, chorine look-alikes, dressed identically and performing in unison, present failed or flawed copies.

Anderson mentions that, when she samples something from a previous time period, this will alter it. The term 'sample' is key here because that is precisely what she does throughout *Yippeee!!!*. For example, various references to Berkeley's title number in *Yippeee!!!*'s '42nd Street' section are evident, but they are small, sampled fragments that are reassembled differently. Recognisable movement material includes: shimmying shoulders; waving a hand as if greeting; jumping up and down vertically on two feet; and kicking the legs to the side. However, whereas these moves occur in Berkeley's number only a few times, interspersed with many other stylised street activities of passers-by, stall vendors and so on, in Anderson's choreography these samples are repeated incessantly, becoming multiples. Extracting only these fragments and continually repeating them is another example of hyperbolic irony discussed above. They are further instances of failed copies that are also mutations.

Copying from film to dance inevitably involves transfers of material from one medium to another. As Anderson points out 'if you reconstruct something from a film it won't all be present and it won't make sense – because of the frame whole chunks will be missing and you have to solve problems and some things won't work. We tried to be accurate but it's impossible, so we filled in the gaps' (in Hargreaves 2007: 25). These gaps were filled in various ways. For example, some of the classic Berkeley overhead shots, where women's bodies are framed to form kaleidoscopic patterns, were replicated in *Yippeee!!!*. Four dancers posed together, one standing, another lying on the floor underneath facing up and the other two kneeling at the side to form a diamond, moving their gloved hands together in and out of the centre to make intricate flower-like patterns with their fingers. They presented to us, vertically, the horizontal ever-moving, opening, turning and closing flower petal patterns that Berkeley's camera had looked down on. The suggestion of pollens and stamens here is typical of Berkeley's sequences which are often implicitly reproductive. Throughout *Yippeee!!!*, Anderson and Vincenzi hyperbolise and make this explicit in a perverse manner. In this flower image, the transfer in scale from bodies to fingers not only underlines the ways women performers are belittled in Berkeley's films, but it is also an instance of mutations emerging from failed copies. The four dancers rotate their formation to the side after each flower sequence, rolling over and changing places, mutating each copy as different dancers' hands perform new roles in the sequence. Mutations spread like viruses and are rife in *Yippeee!!!*, they creep up on us in an uncomfortably surprising and shocking manner.

Patterns and mutations

Anderson's interest in patterns is also evident in virtually all of the ensemble numbers in *Yippeee!!!* with dancers entering, traversing and exiting the space in lines, side by side or following each other. These repeat ad infinitum giving the impression that whilst the performers are 'onstage' they are programmed to repeatedly follow one another, almost as if they cannot stop. Straight lines curve, meander and snake around each other, forming new shapes and new groupings. Repetition produces multiple copies, establishing patterns which then evolve and kaleidoscopically form yet more patterns. As the patterns replicate each other, they gradually spread and mutate, changing from one form to another. They become invasive. Mutations also occur within and between individual performers, who change appearance with each costume, adding and losing limbs, protuberances and appendages. They grow dangling, penis-like snouts, third legs at their hips, and sperm-like tails. Body parts also disappear into costumes such that performers occasionally appear armless or headless. Reproduction on several levels is rife. Sexual reproduction is ridiculed in the sperm dancing. More disturbingly, reproduction of viruses and suggestions of cloning are hinted at in the mutations, and strange hybrids created. Having said this, the mutations suggested are not always negative. Anderson informed us that the third legs some dancers appear to grow were 'also a suggestion of mutation – maybe of evolution into a more efficient dancer with more legs' (Anderson 2008). The repeated patterns and multiples, which pervade the show, continually reinforce this theme.

Patterns and transgressions are also reflected in and magnified by the repetitive rhythms of Steve Blake's musical score, which draws heavily on a range of popular music sources. At times electronic distortions make the accompaniment sound decadent and mechanical as if emanating from a fairground, a pinball machine or barrel organ. The music, like the dance, seems to have 'on-stage' and 'off-stage' qualities. During the big on-stage numbers, straightforward rock music, with repetitive, pattern-forming melodies, echoes and supports the performance. Blake, who has been collaborating with Lea Anderson since the 1980s, knows exactly how to give dancers the musical support they need with exquisitely judged, musical redundancy. During the 'off-stage' interludes, however, the music becomes increasingly experimental, resembling avant-garde jazz. Occasionally in these interludes the music appears to be slowly going out of tune; out of its normal range. Notes sound as if they are surreptitiously transgressing, going where they shouldn't, as the companies' website suggests when it describes the music as 'virulently infectious Electro-squeak groove' (Cholmondeleys 2006). The music is also mutating because the sounds gradually distort like the bodies of performers.

Hannah Höch

Multiplicity, perverse hybridities and ironic humour are also apparent in another of Anderson's sources: the photomontages of the Weimar, German Dadaist, Hannah Höch. Many of these were created at the same time as Berkeley's films. In fact, Berkeley (1895–1976) and Höch (1889–1978) lived through the same historical events. Both shared similar fascinations with the possibilities for pattern-making of multiple images of the female form. Her photomontages and his dance numbers often focus on chorus girls' long legs and single-strap dance shoes. Like Anderson, however, her interest was not uncritical so that it is not surprising that Höch's work became a source for *Yippeee!!!*. Both artists share a fascination with the mass media, Hollywood films and fashion magazines.

Another shared concern is androgyny, and the ambiguities of gender and sexuality. Anderson told us that one Höch work which they frequently consulted was *Dompteuse* (1930). This work assembles a female mannequin's head, masculine muscular, hairy arms, and a flat-chested torso, with a close-fitting skirt to make up a seated, androgynous figure. This image has been torn out and superimposed on a photograph of sheet metal with brass studs. The studs surround the body like a frame but in places the torn edges overlap it. Maud Lavin suggests the tension between the image and the studs signifies 'both containment and overflow, connoting a controlled violence' (1993: 193). There is a similarly contradictory tension in *Yippeee!!!* where numbers are contained on-stage by the set but overflow into the off-stage interludes, suggesting similar connotations of controlled violence. Anderson and Höch have both created scrapbooks of material from magazines, newspapers and fine art as part of their working process. They have similarly ambivalent responses to such images both celebrating and critiquing them. Lavin suggests that Höch's photomontages employ 'pleasure in the aesthetic aspects of media photographs mixed with an angry critique of certain social stereotypes' (ibid.: 124). Anderson has a similarly ambivalent relationship with her material. Her often expressed obsession with choreographing patterns means she has always been fascinated by the kind of symmetry and unison that Berkeley's work exemplifies. Her choreographic approach in *Yippeee!!!*, however, has more in common with Höch's practice of appropriating and fragmenting imagery.

Höch assembles body parts from different sources to construct new images where the parts, because they are from differently gendered and sized originals, do not fit 'naturally' together. Her collages are full of hybrid bodies with outsized heads and very small limbs, with extra appendages or missing components and differently gendered parts (see Figure 9). The dancers' faces in *Yippeee!!!* (see Figure 8 above) also resemble imagery from works such as Höch's *Happy Lady* (1923), where a large grinning mouth has been placed on a much smaller face. Another example of a distortion in scale in *Yippeee!!!* which looks like it originated from a Höch work occurs during the 'hand jive' section (see Figure 10). Two performers flank a third, holding up a white disk containing a magnifying glass pane through which the central performer's face is seen enlarged. The effect is uncannily disorienting, because we only see the central performer's supersized eyes, nose and mouth, which appear disembodied, floating above their arms and legs. As in the flower petal hand dance described earlier, the central performer's hands create patterns that recall Berkeley numbers, with again a disjunctive change of scale. One hand beckons seductively but then turns over and, with undulating fingers, performs a cross between a stroking gesture and a sign to keep one's distance. Teasingly oscillating between saying 'come hither' or 'hold back', this hand jive is cheekily ambivalent. Then both hands form a gaping enclosure that rotates with fingers unfurling open before closing shut. The whole hand jive is repeated with the three dancers moving sideways and, as they exit, they remove the frame so the audience can see how the magic works. Lavin suggests that Höch 'deconstructed melodrama by magnifying and exaggerating features in her portraits' (1993: 125). Anderson is doing something comparable with similarly disorienting and disjunctive effects.

The critical elements in both Höch's and Anderson's work are evident in the ways in which they reveal the constructed nature of much of the imagery they also celebrate. Both Höch and Anderson expose how such idealised images are produced by showing the joins. Höch does this by cutting up her source material and reassembling it in photomontages where the cut edges are clearly visible. Anderson does it by revealing offstage activity and failed copies. Both artists provocatively defamiliarise their original source material and recombine it with alien material that doesn't quite fit, pointing up and often ridiculing the original. In both Anderson and Höch's work, signs become partially disconnected from the discourses that would otherwise give them their meanings so that they still retain a trace of their 'normal' context, but make this appear uncanny. It is this uncanniness within Anderson's appropriations from Berkeley that some of the London critics found disturbing. These partially disconnected signs are particularly evident in the five duets in *Yippeee!!!* that come between the large set pieces and contain some of Anderson's richest and most complex choreography.

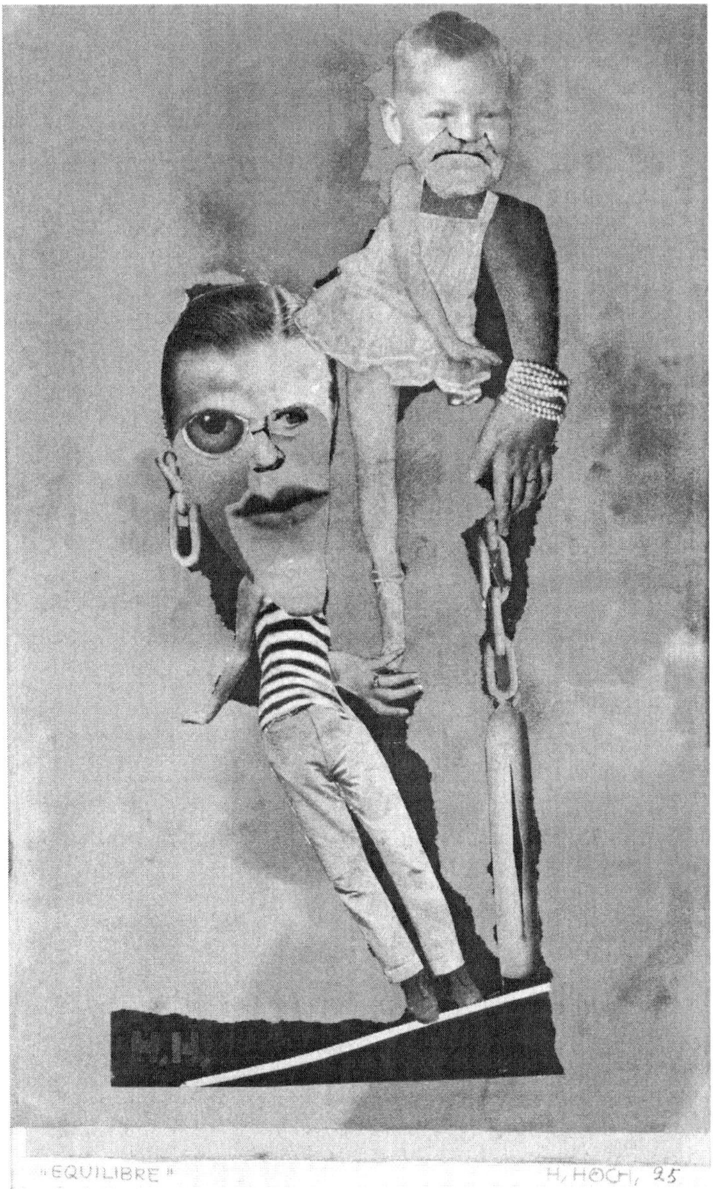

Figure 9 *Equilibrium*, 1925 (water colour and gouache on paper), by Hannah Höch (1889–1978)

Private Collection/ The Bridgeman Art Library.

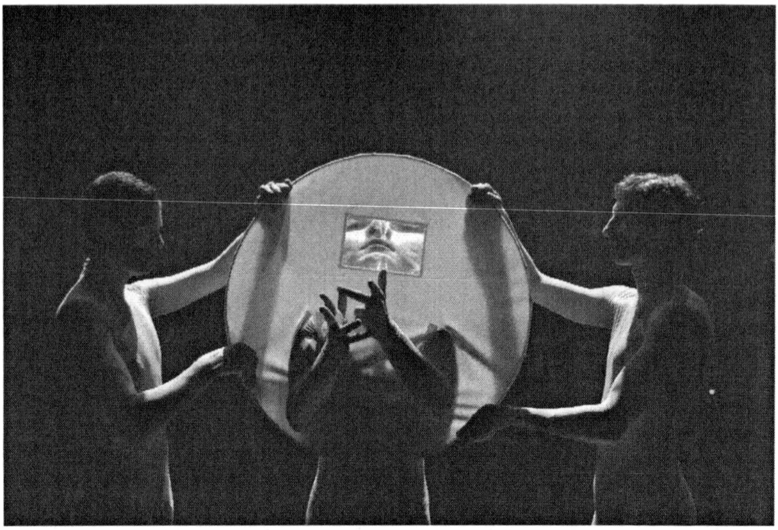

Figure 10 Lea Anderson's *Yippeee!!!*: hand jive
Photo: Pau Ros.

Duets

The musical accompaniment for these duets consists of a simple melodic improvisation over a recording of percussive tap dancing. Each accompaniment is slightly different. In the first, the taps are slow with only occasional piano jazz chords, while the second is a little more upbeat but with a dreamily meandering tune on an electric organ. The dancers' movements, however, rarely appear to have any direct relation to the rhythm of the recorded tapping. At first glance, their legs and feet seem to be performing moves from the jazz dancing of Berkeley and Höch's day, but their steps aren't actually making any sound, and their torsos and arms seem to be making shapes that suggest entirely different kinds of movement vocabularies. Spines and backs sometimes bend and contract in ways that recall German or US modern dance from the 1920s and 1930s; or a dancer may point a forefinger off-stage, or at their partner, or at their own toes as they kick forwards, or at their own temples as if to say 'silly me'. The two partners generally keep close to one another throughout their duet, frequently touching or manipulating one another in ways that alternate between intimacy and impersonality. Sometimes they embrace or nuzzle up against one another. At other times one steps over the other one's outstretched limb as if surprised to find it there. Sometimes they are as closely synchronised as Astaire and Rogers, and then they become randomly disconnected, facing away from each other in opposite directions although performing the same step. The strangeness is exacerbated by costume accessories. In the first duet, a red cloth patch is strapped over each dancer's nose, almost like a medieval torture device. In the fifth duet the dancers each have a red velvet vagina covering their mouth and nose. Often a duet that is full of delicate and precise details degenerates into grotesque farce. At one point dancers start to prance like gawky children; at another, they bend over one another, hanging on to the nearest body part while awkwardly waltzing like clumsy ogres.

The duets, as a whole, seem to keep hinting at familiar ways of moving – jazz dancing, modern dance, comedy sketches or movements from ordinary, everyday behaviour. Overall, there is very little repeated material and the type of movement keeps changing. Because of their density and fragmentation, the duets seem frustratingly familiar but at the same time elusively unreadable. Anderson told Hargreaves that the material for the duets came from one of Laurel and Hardy's best known dance routines in their film *Way Out West* (1937). She worked with the dancers, she said, on splicing this with Hannah Höch pictures. While writing this essay, Ramsay viewed film of duets together with the sequence from *Way Out West* but couldn't see any connection between them. When we queried this, Anderson told us 'we sampled this [Laurel and Hardy's dance] in its entirety and then replaced every move with something inspired by a Hannah Höch collage' (Anderson 2008). The end result seems, to us, to look as if the dancers are trying to perform something they don't really 'understand'.

In his 1985 book *The Man Who Mistook His Wife for a Hat*, Oliver Sachs gives an account of the perceptual worldview of patients suffering from the severest of global aphasia. Although these patients are intelligent, their condition renders them incapable of understanding words as such. Because of this inability, Sachs points out, aphasiacs become extremely sensitive to what he calls the feeling tone of expression. He suggests one cannot lie to an aphasiac:

> He [*sic.*] cannot grasp your words, and so cannot be deceived by them; but what he grasps he grasps with infallible precision, namely the expression that goes with the words [...] Aphasiacs are preternaturally sensitive to any falsity or impropriety in bodily appearance or posture.
>
> (1985: 78)

In *Yippeee!!!*, it is as if both the dancers and their audience are suffering from a special kind of aphasia which renders them incapable of understanding the cultural codes and discourses that ordinarily make dance movement meaningful. While *Yippeee!!!*'s humour sometimes depends on some basic familiarity with the kinds of dances Berkeley created, it doesn't matter whether or not beholders pick up on particular references to specific film numbers; nor do they need to know anything about Hannah Höch, or Laurel and Hardy, or any of the other sources which Anderson and the dancers drew on while creating the piece. Because these sources have been defamiliarised through disorienting appropriations and fragmentation, any meanings they might ordinarily have had, are turned into ghostly memories. In their absence, the beholder becomes particularly sensitive to what Sachs calls the feeling tone of these movements, which thus reverberates with incremental intensity.

Conclusion: 'Don't mention Busby Berkeley'

Yippeee!!!, like many of Anderson's works, had two tours, the first following the London premiere in the autumn and a second follow-up tour the following spring. Because some of the critics had responded negatively to the piece's references to Busby Berkeley, promoters external to the company advised the theatres taking *Yippeee!!!* not to mention Berkeley in their pre-show publicity. For the spring tour a new poster was designed. Whereas the original poster had a Hollywood, art deco theme and showed a Berkeleyesque chorus line, the new poster played down these aspects, using instead a collage effect that recalls Höch's work. The warning issued to touring theatres not to mention Berkeley seems bizarre, given how extensively Anderson and the dancers mined his films while making *Yippeee!!!*. Read against the grain, the problem seems not to have been the fact that *Yippeee!!!* drew heavily on Berkeley's films, but that its references were not straightforwardly celebratory. The problem was rather what *Yippeee!!!* implied about Berkeley: that he was misogynist and sadistic, that there was a queer, fetishistic underside to many of his dance numbers. *Yippeee!!!* also suggested, however, that, despite all this or maybe even because of it, particularly from the point of view of a female choreographer at the start of the twenty-first century, Berkeley's work can be simultaneously fascinating and repulsive. *Yippeee!!!* proposes looking again at Berkeley's work as if one neither knew nor understood anything about it or, as we have suggested, as if one suffered from a special kind of dancing aphasia. According to

Sacks this renders one particularly sensitive to falsity and impropriety in bodily appearance and posture. Anderson's dancers show this kind of sensitivity through the hilarious way they relish every opportunity the choreography offers them for mimicry and perversion. In fact *Yippeee!!!*'s response to the contradictory pull of fascination and repulsion is laughter. Laughter is uncontrollable and infectious, a flow of intense positive energy. To say 'Don't mention Busby Berkeley' is a defensive reaction, an attempt to contain something that is worrying. It is a response based on fear of the danger of contamination. Ironically, this is fear of the kinds of mutations with which *Yippeee!!!* abounds.

Propelled by the infectious, rhythmic patterns of dance and music, and enhanced by countless costume changes, mutations spread invasively throughout *Yippeee!!!*. They are evident in the performance's transgression of normative on-stage/off-stage divisions and in its suggestion that gender and sexuality are fluid and fluctuating. Within this world which *Yippeee!!!* creates, the hybrids that emerge are positive and welcome. For those who find them disturbing and unsettling, *Yippeee!!!* offers only the laughter of derision. Mutations embrace potential as a way of tapping new sources of energy for finding alternatives to present ills. One source of this is in the multiples which destroy the idea of unique, original, authentic origin by exploding the shell of individuality in order to release intensities. Thus, whereas Busby Berkeley's perfect patterns reinforce unequal power relations, *Yippeee!!!*'s multiples suggest non-hierarchical ways of sharing energy. In our reading, the critical reception of *Yippeee!!!* is an instance where a number of factors conspire together to try to tame and constrain what is uncontrollable, thus limiting this potential. At a time when, as the wittily ironic performances in *Yippeee!!!* point up, and Anderson indicated, there is little to say 'Yippeee!' about, we contend it is crucial to keep open this potential for scandalous alternatives.

7
Corporeality and Materiality in Pina Bausch's Tanztheater: Notions of the Irreparable

Valerie A. Briginshaw

The repeated use by many critics of the words 'visceral' and 'intense' or 'intensity' was notable in the reviews of the 2008 visit of Pina Bausch's Wuppertal Tanztheater to Sadler's Wells Theatre, where the company performed *Café Müller* and *The Rite of Spring*. For me, these are 'catch all' terms that point to the importance of the materiality of Bausch's dancers' bodies, but to a certain extent they hide as much as they reveal. In this chapter I intend to explore the ways in which Bausch's works mine corporeal materiality and in the process expose the complexities that surround these notions.

In Bausch's pieces, her dancers continually foreground the body itself, its corporeality – its various parts, the ways in which they work or don't work, what they can or cannot do – and the body's materiality – what it is made of, how it behaves. Relationships between corporeality and materiality are revealed as open and complex. The similarities and differences between what a body is and can do, and what constitutes it or makes it up are exposed in Bausch's works as not straightforward and deterministic, as our cultural codes imply, but as indeterminable and rife with multiple possibilities. Reciprocal implications in the gap between corporeality and materiality, which language has created, are revealed by Bausch's dancers' performances. Her dancers' bodies, as they foreground their corporeal materiality, are always open. In their openness, they are fragmented. They reveal cracks and fissures. They do not portray the contained, normative bodies that we have come to expect from dancers, but rather present flawed and failed bodies. In their uncontrollable outpourings of excessive and surplus energy, they appear irreparable, without remedy.

The Italian philosopher, Giorgio Agamben, coins the notion of 'the irreparable' to explore further the relationship between essence and existence, in his terms, 'the old problem of metaphysics', dealt with in Wittgenstein's *Tractatus* and Heidegger's *Being and Time* (1993: 89). Agamben's work is particularly pertinent for an exploration of Bausch's dances, because he, like Bausch, recognises the limitations of Western thought, discourse and language. The Western metaphysical tradition in his view 'has hastily left aside' (ibid.) these important issues. Bausch is notoriously reluctant to talk about her work because of the propensity for words to limit and close, thus denying the openness and potential for multiple meanings that her dances provide. She has said of her work, 'Words are too small for me to describe this process' and 'you can't formulate this because it really belongs to another language altogether. It is all about this other language, which with words you can only try to come close to' (1990, Bausch interviewed in Kirchman TV documentary, translated by Thomas Kampe). Through examining three female solos from Bausch's works: *Walzer* (1982), *Palermo, Palermo* (1990) and *Der Fensterputzer* (1997), alongside Agamben's writing on the irreparable, I hope to come sufficiently close to this 'other language' that Bausch mentions, to indicate the implications of her work for relationships between not only corporeality and materiality but also essence and existence.

Walzer – Jo Ann Endicott

In Jo Ann Endicott's lengthy solo in *Waltzer* (1982), in a blue swimsuit and black high heeled shoes, she aggressively confronts the audience with her corporeality, with her bodiliness. She begins by staring fiercely at her co-performers on stage, and at the audience. Discussing the solo later, she asserts: 'I looked at the others for a long time, eye ball to eye ball. I had lots of time. I was feeling super hard. I looked at them at length until I made them lower their gaze one by one. Look! The spectators sitting on the stage were also wrapped in this look' (Endicott 1999: 20). These determined looks seemed to further emphasise her corporeality. Explaining the origins of the 15-minute solo in her responses to questions or statements put to her by Bausch, Endicott indicates that the first statement was 'depreciation/debasement', and that her reply 'became a solo in the piece' (ibid.). Bausch's method of creating dance material is to give her dancers open-ended questions, statements or tasks to explore by delving into their own backgrounds, memories, experiences. The dancers are encouraged to be open, associative or intuitive in their performative responses rather than literal and closed. Endicott's bodily reaction to 'depreciation/debasement' resulted in an exploration of her corporeality in terms of what her body could and could not do, including its considerable balletic skills. Endicott, in a later interview, describes ballet classes she took when younger, asserting: 'I had talent, knew how to do things with my body and was always top of the class', but in the Australian Ballet company she was 'forever told: your face is too fat' and she admits 'For classical ballet, I was simply a couple of kilos overweight; I had this round face, and didn't look like any of those around me' (1999: 51).

After marching noisily into the auditorium up the side aisle and calling for light, in her own words, 'in a commanding tone'. 'I want more lights: two spotlights, I said two!', she shouts. She returns to the stage dragging on a large table and shouting, 'I don't need your help or anybody else's help. Thank you!' She repeatedly asserts this whilst sitting on the floor drawing a chalk outline around her legs, committing her body shape to the stage, and simultaneously biting bits off an apple which she spits out violently. Sitting on a chair at the front of the stage facing the audience, she addresses them directly saying, 'when you sit like this', slumping, letting her legs fall open and pointing to her thighs, 'do your legs look fat, ugly and revolting?' Putting her hands under her breasts and pushing them up she spits out 'and when you wear a bra your boobs are up nice and high, and if you don't', dropping her hands, 'they sag'. In her direct address to us we can identify with this, we too have 'imperfect' bodies that behave like this. As Agamben indicates when discussing the irreparable: 'the world of the good and that of the evil contain the same states of things [...] they are perfectly identical [...] The one who is saved and the one who is lost have the same arms and legs' (1993: 92).

Endicott then, standing up and strutting about the stage, rapidly and sternly, points to various body parts identifying them and what they can do. 'I have long hair', running her hands through it, 'I can wear it up if I wanna feel like a lady', showing us, 'or down', letting it drop, 'if I wanna feel like myself'. 'Underneath I have a face', pointing to it, 'with two eyes to see with, a nose to smell with, a mouth, teeth, tongue, lips to kiss with, to eat with whatever I like, to speak with, ears to hear with', pointing to each in turn. 'Arms, I can move my arms!', she swings them around, 'hands, I can make a fist, if I want to', demonstrating, 'fingernails, if I want to, I can [...]' she scratches the air in front of her like a snarling, caged animal. Getting up, she continues, 'if I stand up like this', stretching up, 'it looks fine, if not', slumping, 'it's a slob', she spits out, 'yes a slob, a slob!', she emphasises, screwing up her face, implying disgust. 'And behind I have a back', she says, proudly turning and pointing to it. 'I worked, not months, but years and years and years and years for this back, stretching it out, getting that spine up straight, for this back to stand up proud', she asserts, stretching up tall and crisply drawing a straight vertical line with her hands in front of her, indicating her upright spine and the effort required to produce it. 'I have got two long legs and feet to stand on', she declares, and walking, she shouts, 'and if I want to, I can walk!', and louder,' I can walk!', striding about the stage.

She hurls her declarations and demonstrations of her bodily competences and limits at the audience with such vehemence that the forcefulness of her 'full-on' corporeality cannot be ignored. Agamben suggests that 'irreparable means that [...] things are consigned without remedy to their being-thus, that they are precisely and only their *thus* [...] for them there is literally no shelter possible [...] in their being-thus they are absolutely exposed, absolutely abandoned' (1993: 39). In Endicott's forceful, confrontational bodily performance she is exposed and perhaps also abandoned, abandoned by those ballet teachers from her past who told her, her face was 'too fat'. In this revealing solo, Endicott presents us with things (in this case her body) *just as they are*, which is how Agamben describes the irreparable (ibid.: 89). There is a strong sense, we feel, after witnessing this performance, that things cannot be otherwise.

Sitting down, Endicott puts a sweet in her mouth and after chewing it, removes it and sticks it on her knee from where she picks it up with her mouth, chews again and removing a shoe, sticks the sweet on her bare heel. She then stabs it with the shoe's stiletto heel and returns it to her mouth. Her defiant performance of these bodily party tricks, like many episodes in Bausch works, contravenes expected norms of bodily behaviour. This misbehaviour is humorously shocking, like Endicott's earlier spitting out of half-bitten pieces of apple, it blurs bodily boundaries between inside and outside. It further emphasises the openness of bodies in Bausch's work; the ways in which they exceed their boundaries. What is normally kept inside comes out, interrupting and disrupting any attempts at clear articulation. These unexpected eruptions are out of place; they daringly go beyond the bounds of what is expected of bodily behaviour. Endicott's performance seems to be veering between, on the one hand, a confident display of bodily accomplishments, and on the other, a desperate, almost childish, desire for attention; to be noticed. Agamben explains that 'According to Spinoza, the two forms of the irreparable, confidence or safety (*securitas*) and despair (*desperatio*), are identical' and 'every cause of doubt has been removed [...] things are certainly and definitely thus' (1993: 91). Endicott's irreparable performance allows for the coexistence of what might be seen, in the limiting terms of first philosophy (Agamben), or in words (Bausch), as complete opposites – confidence and safety on the one hand, and despair on the other. It dismantles the boundaries, of not only the inside and outside of the body, but also between forms of behaviour normally thought of as opposites.

The next stage of Endicott's performance involves her demonstrating a series of ballet exercises; stretches using first a chair and then the piano as a ballet barre, where she swings her leg way up in front and behind her, and then ends dramatically on the floor in the splits. Throwing her body into these actions, she appears on the brink of losing her balance, but she miraculously maintains it. After each stunt she lets out a loud half sigh, half groan, revealing the effort required that is normally hidden: another display of the irreparable where confidence and despair coexist, where, as Agamben affirms, 'being that is irreparably thus *is* its *thus*; it is its only mode of being. (The thus is not an essence that determines an existence, but it finds its essence in its own being-thus, in its being its own determination)' (1993: 93). Endicott has revealed an open corporeality – a body that cannot be otherwise, only thus. As Agamben indicates, in one sense it cannot be described in words since the being-thus of this irreparable body is not an essence that can determine an existence that follows from it, its essence and existence are rather bound up in its being-thus. As Bausch herself claims of her work in general, 'it's not something you can do with your intellect [...] it can only work if we avoid anything explicit [...] if we avoid this [...] I think there is the possibility for another kind of language' (interviewed by Meisner 1992: 15).

Palermo, Palermo – Julie Shanahan

This other kind of language is rooted in the body. Whilst in Endicott's *Walzer* solo the body's corporeality is foregrounded, we are graphically shown what the body can and cannot do. We are also shown the body's materiality. What it is made of, how it behaves, its properties, its insides and outsides are revealed as part of this. In Julie Shanahan's solo that opens *Palermo, Palermo*, the materiality of the body is further exposed and explored.

Shanahan, in high heels and a knee-length shift dress of patterned silk, falling off one shoulder, picks her way across bricks and rubble from the fallen wall whose collapse marked the beginning of the show.[1] Facing the audience, she draws a large X on the floor with chalk, stands on it, chucks the chalk away, and then draws a black X on her face literally crossing it out, obliterating it. Marking her place on stage and herself with the same cross sign, she seems to be consigning herself to some sort of ritual display object. We hear recorded piano music introducing a women's blues song 'Why don't you do right?', performed by the Mean Mothers. As if the music has brought her to life, Shanahan slowly raises her elbows, placing her hands on the back of her head, framing her face, like a model on the cover of a fashion magazine. She assumes a typical statuesque, mannequin-like pose, one leg slightly bent in front of the other. A husky woman's voice sings, 'you had plenty money in 1922, but you let other women make a fool of you' and as Shanahan reaches her classic model pose, we hear the song's title line 'why don't you do that right?', like an inner voice chastising and questioning her as she poses.

With the next verse of the song, she begins a gestural dance, smoothly undulating her hands in time to the music, putting a hand on her crotch, scrunching up her dress and making a gesture as if gouging out her innards as she pulls her dress tightly round behind her emphasising her statuesque figure. Two men come to either side of her, hold her by the elbows and waist, and gently lower her face-down, like a shop window dummy, to the floor. As they disappear, she calls, 'Pick me up'. They return her to standing in one piece, and walk off. She calls one back 'Thomas, take my hands'. When he does so, she immediately wrenches her hands away to resume her pose. She repeats her request and her action twice, each time getting more agitated and raising her voice. Next she requests, 'Hug me', but when he does, she breaks free, then repeats 'Hug me Thomas', then 'Harder!', then more loudly she implores 'Harder!', but almost immediately pushes him away screaming, 'Go!'. She is passed a pile of black earth on the shallow lid of a cardboard box, which she throws up above her, falling to the ground as it lands on and around her, sticking to parts of her dress. Sitting up, Shanahan repeats the earlier gestural hand dance to the music before lying down flat again in her model-like pose. As the two men lift her back up, she pulls her dress behind her at the waist re-emphasising her svelte proportions. Then sitting on a chair, keeping her hands behind her head, she repeats her terse instructions 'Take my hand, Mark!', immediately returning it as before, and again, 'Take it!'. After repeating orders to 'Hug me!', each time breaking away, she asks the two men to throw rotten tomatoes at her. When they do, she remonstrates with them 'Throw them at my face!' and then almost immediately, 'No, stop!'. Then again, almost at the end of her tether, she screams, 'Throw them in my face!' whilst screwing up her face as they hit it, and then she asks them to stop. By now, she is covered in dirt and bloody red tomatoes dripping off her and her dress. (see Figure 11). Her body seems irreparable. She appears 'consigned without remedy to [...] being-thus'. In Agamben's terms there is literally no shelter possible, she is absolutely exposed and she appears absolutely abandoned (1993: 39). Her exposure is paramount.

Agamben, discussing the irreparable, suggests that 'Existence as exposure is the being-*as* of a *such*'. He continues: 'To exist means to take on qualities, to submit to the torment of being such (*inqualieren*). Hence quality, the being-such of each thing, is its torture and its source – its limit. How you are – your face – is your torture and your source' (1993: 97). Shanahan's performance, her exposure, can be seen as the 'being-*as* of a *such*' or in Bausch's terms, when describing the affects of her work, 'it is a feeling – but a feeling which is very precise; it is not just some vague feeling, it is something absolutely precise' (in Meisner 1992: 16). It cannot be otherwise.

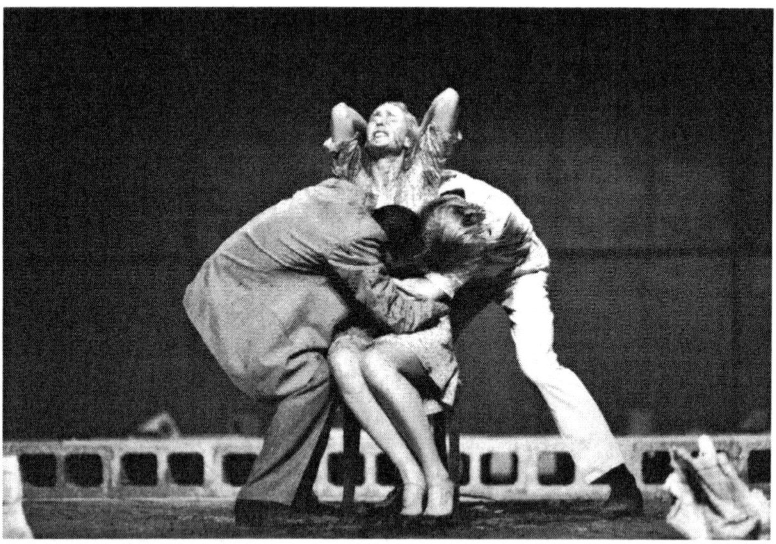

Figure 11 *Palermo, Palermo* Choreography: Pina Bausch, Tanztheater Wuppertal, Dancer: Julie Shanahan
Photo: Matthias Zölle.

In this solo Shanahan continually contradicts herself: 'hug me', 'let me go', 'throw', 'stop' – she is a split, schizophrenic subject who defies any kind of unity. She reveals a marked, fractured body with open wounds, a ruptured subjectivity, falling apart at the seams. To a certain extent, it is as if she is lost, as if she loses herself in this solo. In his discussion of the irreparable, Agamben claims that 'losing oneself to the point of not being able to conceive of anything but things, and only then, in the experience of the irremediable thingness of the world, bumping into a limit, touching it. (This is the meaning of the word "exposure")' (1993: 102).

Shanahan's seemingly schizophrenic need to be loved (hugged) and despised (pelted with rotten tomatoes) simultaneously, and Endicott's veering between supreme confidence and utter despair, also simultaneously suggest the reciprocal implications that exist between corporeality and materiality. Similarly, Shanahan's marked body, with its crossed-out face and stained dress, and the dirt and spillage of tomatoes on the floor around it, reveals graphically the ways in which dancers change material and material changes them, the ways in which materiality and corporeality mark each other. This is vividly apparent in many of Bausch's works where her dancers in various states of undress or disarray, their bodies exceeding their costumes, their flesh exposed, perform on or in: peat/earth (*Rite of Spring*); leaves (*Bluebeard*); water (*Arien*); grass (*1980*); carnations (*Nelken*); and red silk petals (*Der Fentsterputzer*), for example. The smells and sounds of these elements, and the mess they make of the performers' often sweaty bodies, reinforce and remind us of the materiality of our own bodies. In Bausch's film *Lament of the Empress*, performers wrestle with and in mud, water, wind and a snowy blizzard: the reciprocal relations between materiality and corporeality are again immediately evident. Endicott's chalked body outline on stage, her remnants of apple strewn about and the openness of her solo generally are further evidence of not only the foregrounding of corporeality and materiality but the reciprocal relations between them.

Endicott's and Shanahan's solos are typical Bauschian performances which, through their irreparability, through their being 'thus' or 'such as', reveal a potentiality to not not-be (Agamben 1993: 105). They cannot not do what they do, they cannot be otherwise. Agamben explores in some detail the important significance of the words 'thus', 'such as', and 'as', the 'as' in 'as such' or 'such as'; the how it is and what it is; the how it exists as it does. He declares: 'Language says something as something: the tree as "tree", the house as "house". Thought has been concentrated either on the first something (existence, that something is) or on the second (essence, what something is), either on their identity or on their difference. But what really has to be thought – the word "as", *the relation of exposure* – has remained unthought' (ibid.: 97–8, my emphasis). I am suggesting that Bausch's dancers' bodies, in their irreparable exposure, are attempting to think the 'as', that has previously been unthought, which, Agamben argues, hovers between existence and essence. Rather than thinking existence or essence, identity or difference, the 'as' in 'as such' or 'such as', through the relation of exposure, is revealed in performance in the reciprocity of the relations between materiality and corporeality.

What we are grappling with here is bound up in the grammar we use and how that determines, limits and constructs the way we think and conceive of things. Both Bausch and Agamben, in their very different ways, are trying to get beyond the limitations of this grammar, of these thought processes bound by, for example, seeing things in dualisms. Bausch's performances and Agamben's writing both explore the in-between where words such as 'as' and 'thus' operate to articulate relations. The repeated workings of language and discourse, to a certain extent, make us into the people we are. Their reiterative practices subject us. Bausch provides graphic examples in her work of the ways in which these reiterative processes operate.

Der Fensterputzer – Mechthild Großmann

One of these is a monologue by Mechthild Großmann, which is interspersed episodically throughout *Der Fensterputzer.* Großmann, addressing the audience directly with a hand-held microphone, has been recounting elements of her relationship with 'her man'. Whilst licking an ice-cream, she has introduced her story with the words, 'there is a man who loves me, but he's not clear about it, I always have to show him the way to his knowledge.'[2] As Großmann addresses us, she seems to be sharing confidences that suggest this is a long-term relationship. For example, she says, 'when we eat together, I look at him, and then at the plate, and then at him' and 'I often give him presents'. Later she confides, 'only one possibility remains that he won't be afraid of me any more. I have to get him used to me little by little' and 'when we sit in a taxi, I put my arms around him and whisper hot words in his ears. Going round corners I emphasise the vulnerability of my body.'

Her final speech occurs at the end of a male ensemble dance where the dancers behind her are all scuttling across the stage on hands and feet with upturned torsos. They are predominantly young men who exude suppleness in their performance. Großmann has been performing with Bausch since 1979. In 1997 she was almost 50.

Gesturing towards the dancers behind her, she says 'When I was younger I made a mistake [...] you should always do these exercises whatever happens, in all circumstances you have to do them, because your stomach', she says, turning sideways and revealing a bulge tightly constrained by her black, close-fitting dress, 'will be nicer', she pats it with her hand; 'flatter', she slaps it more forcefully; 'harder', this time she really whacks it for emphasis. She continues, 'My husband, he loves my lovely body and my legs', she gestures towards her legs, looks at them and half turns to display them better. 'He was really proud of them, they used to look quite sophisticated', she claims with obvious pride. The tenor of her voice then changes, it is already deep, but it reaches down even further: 'but I had stopped doing these exercises and that was my mistake!', she blurts out, chastising herself. She continues: 'One time when my husband started to kick me here, here', she points to the sides of her legs, 'he only hit the places which didn't hurt so much like my thighs', she claims, 'but I was unhappy, I had stopped doing these exercises and that was my mistake.' This sentence becomes a mantra reiterated throughout her speech, which she uses to punish herself repeatedly like a metaphorical whip goading her neglected body, just as in my other two examples, Endicott and Shanahan incessantly repeat their bodily displays and rituals.

Sharing with us her affective state, Großmann declares, 'my head was like everything was mixed up, muddled, totally muddled. I watched TV a lot, then my husband came and hit me right on my head', she taps her head; 'all the time, right on my head', hitting herself harder; 'right on my head', harder; 'right on my head', harder still. She concludes saying laughingly, 'I was really hurting on the top!' Then she reminds us again, as if trying to teach us a lesson: 'but I stopped doing the exercises and that was my mistake'. The account continues, and there is a sense in which Großmann seems compelled to give an account, to account for herself in an almost confessional manner: 'Then in hospital, I got a new short hair cut, that was my mistake, I shouldn't have done it, I shouldn't have done it, maybe the new short haircut could have saved something, maybe my husband would have been satisfied with a new short cut wife, but I stopped doing these exercises and that was my mistake.' Her final statement further confirms the power operations in this abusive relationship; 'You see, my husband says, you see, I always told you so, you must do these exercises, you absolutely must do these exercises.'

Repetition of gestures and movement material, and of words and phrases is frequent in all Bausch's works. She does not only use repetition for emphasis, she seems to understand the workings of repetition at a deeper level of language, discourse and actions. She sees that repetition is never repetition of the same, that it is open and multiplicitous. It makes things different. By allowing us to see them again, it helps us to see things differently. But it also shows that repetition can work through embedding and embodying gestures, actions, words and phrases, which can ultimately discipline and subject us. In each of these solos the repetition is intensely affective, because of the ways in which gestures reinforce words, actions reinforce phrases, and corporeality and materiality reinforce each other. In each case they present what Agamben calls a 'being-such-as-it-is' (1993: 100), it cannot be otherwise, it cannot not-be.

Agamben, in his discussion of the irreparable, explores the anaphoric or repetitive ways in which the word 'thus' is used, stating: 'as anaphora, the term *thus* refers back to a preceding term, and only through this preceding term does it identify its proper referent' (1993: 93) He continues: 'In anaphora, through reference to a term already mentioned in discourse, this presupposition is posited in relation to language as the subject (*hypokeimenon*) that carries what is said (hence anaphora has provided the model for essence and meaning, the Aristotelian *ti hen einai*). The pronoun [*thus*...] through anaphora, makes [...] being the "subject" of discourse' (ibid.: 94). In other words, our ideas and their potential are shaped through repetitive and reiterative use of the pronoun, in the ways in which we have constructed language to operate. Particular ideas of what things mean are imposed through the language we use, through discourse. On one level this might seem obvious. But Bausch's performances, through their graphic and materially corporeal use of repetition, reveal not only how these discourses work on the body through forms of self-discipline, but also, if seen alongside Agamben's notions of the irreparable, have the potential for something more.

In explaining the irreparable, Agamben asserts: 'we have to conceive of an anaphora that no longer refers back to any meaning or any referent, an absolute *thus* that does not presuppose anything, that is completely exposed' (1993: 94). Returning to the conventional operations of the pronoun 'thus', he explains: 'the originary fracture of being in essence and existence, meaning and denotation is thus expressed in the double meaning of the pronoun, without the relationship between these terms ever coming to light as such. What needs to be conceived [...is] this relation that is neither denotation nor meaning, neither ostension nor anaphora, but rather their reciprocal implication' (ibid.: 94). This relation, this reciprocal implication, is what Endicott, Shanahan and Großmann show in their solos.

By 'being-such', their 'manner of being is not assuming this or that quality, this or that character, virtue or vice, wealth or poverty' (Agamben 1993: 96). Their 'qualities' their 'being-thus are not qualifications of a substance (of a subject) that remains behind them' (ibid.). They are never '*this* or *that,* but always *such, thus* [...] Not possession but limit, not presupposition but exposure' (ibid., his emphasis). This exposure is neither one of their qualities nor other than them, it is 'what happens' to them by being in relation to language, the fact of being-called (ibid.) or subjected. In the immediacy of the dancers' performances, in their raw openness, what some critics call their 'intense, viscerality', what Bausch reveals is, in Agamben's terms, 'an eternal exposure and facticity [...] an eternal sensation' (ibid.: 99). The dancers are, in his terms, beings that are never themselves but only the existent, 'completely and without refuge' and 'the existent no longer refers back to being; it is *in the midst of being,* and being is entirely abandoned in the existent. Without refuge [...] safe in its being irreparable' (ibid., my emphasis). Bausch's performers, by being in the midst of being, demonstrate the potentiality of the reciprocal implications of existing between corporeality and materiality, existence and essence, meaning and denotation.

Notes

1. See Chapters 5 and 8.
2. Bausch's cast of international performers speak a range of languages and sometimes translate their speeches into the language of the country in which they are performing. Großmann, a German actor, performs her speech in English for the Hong Kong performances of *Der Fensterputzer* and in German in Wuppertal. This description is informed by the performance of *Der Fensterputzer* premiered in Hong Kong in 1997 filmed by Peter Lindbergh, and the company's own video recording of a performance in Wuppertal on 10 November 2006.

Section 4

Two Interviews

8

Interview with Peter Pabst, Sadler's Wells, London, 2 February 2002 and Cologne, 22 February 2008

Valerie A. Briginshaw and Ramsay Burt

Introduction

The interview with Peter Pabst, Pina Bausch's scenographer, that follows is an edited version of a public interview Peter kindly agreed to give as part of a Pina Bausch Study Day we organised at Sadler's Wells Theatre on the occasion of the Pina Bausch Tanztheater Wuppertal's visit to London in 2002. The edited transcription was discussed and extended with Peter Pabst when we visited him in Cologne in 2008 and he agreed to answer further questions about his collaborative work with Pina Bausch. Subsequently, he has seen and agreed this final edited version combining the two conversations. Working on this interview with Peter has been fittingly collaborative in more than one sense. This book is essentially about collaboration – collaborative practices and processes – that we have engaged in whilst writing dancing together over several years. Our discussions and correspondence with Peter Pabst have been collaborative in the ways the three of us have worked together on this text and also in the sense that Peter has been interested in the book as a collaborative project and, as is clear below, he also discusses openly his collaborative process working with Pina Bausch. This collaboration that has lasted now for more than a quarter of a century is characterised by generosity, patience and trust. In his discussions with us, Peter underlined these three terms as keys to his working relationship with Pina Bausch. In the interview, the importance of each of these becomes evident. It should also be clear that the significance of this interview for the book goes beyond the transcription below in that it, and our, extended conversation with Peter have also informed and enhanced our writing of the two chapters that focus on Pina Bausch's work – Ramsay's on *Napoli* and *Palermo, Palermo* in Chapter 5 and Valerie's on Corporeality and Materiality in Bausch's Tanztheater Chapter 7.

The interview

Ramsay: I understand that you first became aware of Pina Bausch's work when you were working at the theatre in Bochum, which is not far from Wuppertal, in 1978, when she came with her group of dancers. Did you come across her work earlier than that?

Peter Pabst: This is a short version of how I met Pina. It had something to do with the Director of the Bochum theatre at that time, Peter Zadek. He was a very wonderful theatre director, who was brought up in England. My first engagement was as an assistant designer at the Bochum theatre. I remember I met him in a corridor, and he said to me, 'I've heard there is a curious, interesting new choreographer in Wuppertal'. He was going to see a rehearsal and invited me to come along. This was where I met Pina for the first time. They were in the ballet studio and they were rehearsing or reviving *Der Frühlingsopfer* (*The Rite of Spring* 1975). We watched for 20 minutes and I didn't understand anything and we left. He had better eyes than I did, and had seen enough to invite her to do a production at the Bochum theatre.

Ramsay: Then in 1980, when Rolf Borzik, who was the designer and collaborator with Pina Bausch through the seventies, when he died, you took over the design of *1980* from his ideas.

Peter Pabst: I know sometimes it's written that the designs for *1980* were started by Rolf but this is not true. It was tragic that Rolf had died so young, so we decided to publish all his past drawings in the programme booklet for *1980* and this might have created the misunderstanding. These were drawings of past productions and didn't have anything to do with *1980*. Then there were two other productions – *Bandoneon* (1980) and *Waltzer* (1982) – which Pina did with two different designers. When she called, I was busy with other work so she took on these two designers. Then she asked me a little bit earlier, to design the piece that became *Nelken* (*Carnations*) in 1982. Since then, our collaboration has continued for half a lifetime.

Ramsay: When I heard you speaking at a conference of theatre scholars in Lyons, you spoke about your working relationship with Pina Bausch. You talked about the collaborative process where everybody contributed, and nobody owned particular ideas. Could you say something about that?

Peter Pabst: It's probably a little bit different. I should explain a little bit about the difference when designing for Pina, from designing for theatre, opera or film. First, when she is starting a production there isn't anything like a structure: no text, no idea, no title. None of the things one usually has. Usually there is something to read, to understand or not understand, and discuss for a long time with the director, the dramaturge or actors. But when Pina starts, there is literally nothing. This makes a big difference for a designer, because there is literally nothing you could design for. Pina has a very different approach. She spends a long time – six, seven or eight weeks – putting questions to the dancers. They can reply in whatever way they like: by saying something, by moving, by doing something alone, or playing a little scene which they can do with some colleagues or with the entire company. These six to eight weeks provide a huge collection of material, which Pina writes down, without making any comment, and with an absolutely neutral face. This is so she neither encourages them nor puts them off but gives them a feeling of freedom. These questions she has put, and the response of the dancers, might give an idea about the direction the production will take. Watching as many rehearsals as possible may give me a way in.

After this process, we still don't know what direction the production will take, neither I, nor Pina, although she certainly does know more or feel more. I remember in the beginning I always made the mistake, because, if one has to create something, one is very nervous, one is vulnerable, it's like having an open wound. You want to save yourself, you try to save yourself by having an idea and by knowing what you're going to do. One always does, it's a reflex. In the beginning I always made the mistake of asking her if she knew where it might be going, and for about five years I always got the same answer. 'Well, you know, I'm listening inside myself, but it hasn't come out far enough yet, so I don't know.' Then I stopped asking. Certainly she knows more than I do, but it's just not come out so that you could put it into words, so that it gets a form. It continues to be research, like walking in the fog. I then start to make proposals, or she will ask me things. Then very often, I have perhaps four, five or six totally different designs, and we spend days in discussion about what things would mean, things we've seen, in the material collection and the little scenes. We think about what it would mean if these happen in this world, in this design, and what it would mean if it happened in the dancers' world. So we try to find out about the inner qualities of the different design options, still not knowing what the piece will be.

A set is not meant to just be beautiful on its own. It needs to be an aesthetic or physical world for the dancers to perform in.

If I have an idea, I may put a picture, or photo or drawing on Pina's desk, and if she hasn't responded to it after a day and a half I take it away, because I know it is now an old idea. This is about generosity and about wasting things. You have to have strong confidence in yourself to just waste or give up an idea. There is no contradiction between this confidence and modesty. Young people often try to fight for an idea, and this is such a negative waste of energy. What they should do instead is just forget about it and have another one.

We try to delay the moment of decision about the set as long as possible, to give us more time to know what the piece will be. Then when this decision is made, it marks a change for both of us. For me, because a new period starts, establishing things that have to be done. Usually, this will happen between five and three weeks before the premiere, so really very late. Every normal theatre would say, sorry, don't even come with a sketch. It's too late. It's up to me to create a miracle to get it done. We have this tradition every year where I then come with my little model. There is a meeting of all the directors, technical directors, and heads of the workshop in the theatre, and they all look at it and say, 'no, sorry sir, it's too late now, you won't get this done.' Then we discuss for the entire day until they get tired. The same happens every year. When they leave to go home in the evening, usually two people, who I like very much, stay behind. These two people are the head of the workshops and one of his staff, who I work with very well. The head of the workshops stands up and says: 'well, you know, they start at 7 o'clock in the morning and if I come into the theatre and find that drawing I'm going to start.' This means we are going to do it. They always make a miracle, they always make it work. So this is a little bit of what happens with the technical development.

For Pina, and for the dancers, this point of decision marks a very important change, because until that moment they have rehearsed on a neutral ground – a lino dance floor – and she has seen all the materials against this background. From this moment on, they all see it, or feel it, or think it, in a different world because of the set. Up to this moment the set design has had no life in it. But from this moment on, we start to work together, and see what I can do with it, and what the dancers can do with it. Then life comes into it.

Of course, if one looks at things, all these different scenes and these funny and sometimes very curious and very strange props that are used, nobody could ever think that up. This comes from all the different ideas that the dancers come up with, or is a reflection on an idea that someone has. It's usually very difficult to say where an idea comes from, who's responsible for an idea. You can't separate that out. It's accumulated from everybody who shares the idea. To be clear, there are different responsibilities. So, in so far as the set is all about design, it's mine. I'm responsible for what we have got there, and for seeing that it works. It's not only the title or credit.

My understanding of theatre is that it is a form of art where very many different talents are involved. And in theatre it's not a piece of art, a result. A piece of art is not a scenic design, a splendid scenic design, nor is it a stunning director's idea, nor is it the most beautiful costume, or surprising or amazing acting. It's the whole, it's a performance. That, in my opinion, is a piece of art. So all these separate responsibilities or ideas, do not work alone, they are not theatre. There is one common responsibility and that doesn't take away the special responsibility for everybody to care about their job and to care about how things are done. Beyond that, I think responsibility has to be shared. If it's not, it doesn't really work. One could think that even if a production is not going well and is not interesting, that one could save it by having a wonderful design, or something like that. That can't be done. I've never heard anyone say, 'Well, the production was boring, I fell asleep but the design was wonderful.' It doesn't happen like that. Either it's all good, or it all fails.

Ramsay: When you talked in Lyons about the set for *Palermo, Palermo* (1989), which starts with the whole proscenium arch filled with a wall of bricks that collapses backwards onto the stage, you said something about the origins of the idea in the walls of the rehearsal room. Can you tell us more?

Peter Pabst: Well you know, I'm amazed what I've said. What kind of secrets I've told you in Lyons! I should be careful (laughter). This is an anecdote about where the idea came from. We were sitting in the rehearsal room not knowing what we were going to do – this state of agony that we always go through. This studio is an old cinema building. In the 1950s they had pleated plastic covering on the walls, and this had rotted so that you could see the bricks of the wall beneath. And Pina looked at it and said it looked like a curtain in front of a wall. After ten minutes of silence I said, well we could have a wall, and she said, what do you mean? So I said, well we could have a curtain and then a brick wall which closed up the proscenium arch. And how do we then get rid of that? I said, well we could knock it down. She said she didn't like the idea of the noise of styro-foam bricks falling, and I realised she thought it would be fake, so I said I'm not talking about a fake wall, I'm thinking about having a real one. Another ten minutes of silence and then she said you're mad. This was how this idea was born. For two weeks or more I tried to get different size bricks and build up little walls myself on the rear stage, walls that were about three meters high that I could build myself and then push down by hand. I was afraid that something would fall the other way. Then of course the city government found out. When we built up the whole wall and reinforced the stage, they said no. There was a whole day of fighting – the unions, the city government, the architects and everybody – until they all became tired and went off. And then it was agreed and my heart was bubbling and what happened when the wall fell was a kind of earthquake, and it really is an earthquake every time it falls.

Valerie: Peter, can I ask you something about the back projected films in *Masurca Fogo* (1998). [Tanztheater Wuppertal were performing this piece at Sadler's Wells Theatre at the time of the interview.] In your earlier works with Pina Bausch, film wasn't used in that way. Did that just come about as an idea at one particular time, and then has been used more? Or have individual pieces suggested it in some way?

Peter Pabst: Well, there are different reasons. If I may correct you, what I'm doing in these pieces is not back projected, it's front projected so it's just inside the room. What I'm interested in, partly, for example, in *Masurca Fogo*, are strong images and what they do to the space and to the performers. In this case, there is a rather claustrophobic space. It's closed everywhere, there's a ceiling, and only one entrance upstage right and that's all. I wanted to see how this closed space opens, very widely with all these images of sea, and how it moves. Sometimes it starts to move around, so this is why front projection is necessary. There are different reasons why it came about. Video has become so cheap and so affordable. This is an advantage, and a danger too, because if you think six, eight years back, you didn't even have the projectors. It was impossible to project a video with this kind of luminosity. Secondly, to do a video, you required huge cameras and a truckload of extremely expensive equipment. This has changed immensely in the past six to eight years. Now all you need is a little box, which costs maybe £800, and you can do the most wonderful things, technical tricks and so on. It's amazing that you can blow a video image up to this enormous size. This is one answer, the most simple one. It's affordable. But this is a danger also because now everybody uses it, not only in the theatre, but in big contemporary art exhibitions. The danger is that if it's used too much, it's overused.

There was a special reason when we first tried. Pina's a person who likes to have everything. At one point there was the question can't we have more? But I cannot build three huge sets and have them change during one performance, it's impossible. So I remember asking, (after timidly trying to hide because I thought I might be hit) if she'd ever have thought about projections. But I wasn't hit, so I thought yes, and this was when I started. We did a co-production *Tanzabend II* (1991) with the city of Madrid and there I did something that I had dreamt of for a very long time, a winter landscape. I put ten tons of salt on the stage. This was very beautiful but very difficult for the dancers, because it's hard for them, even harder than walking on a soft sandy beach. But, on the other hand, it looks very beautiful. It is like fresh snow, it makes the same little noise when you walk in it. I also discovered this was a perfect surface for projection, because it's like a crystal screen, which is usually used in cinema. It reflects in the same way, so putting an image, a slide of the Sahara desert in there, it, all of a sudden, was sand, there was no salt anymore. This could change immensely and entirely the character, this was the beginning. But this isn't necessarily the direction we are going in now. In a recent piece we did, *Água* (2001), a co-production with Brazil, I think there's about two and a half hours projection in that, and about five minutes without projection, this is one extreme. But in *Wiesenland* (2000), the piece which we did before, a co-production with the city of Budapest, there are no projections at all. So this varies. There is never a system behind the use of projections.

I always have my little camera with me and I don't take pictures like a project. I take a picture if I see something interesting and this gives me a collection of material, an archive of images. Sometimes I take an image out of this archive and put it on stage and see what happens. And then sometimes I feel it needs something more and I film something else.

Valerie: Linked to that, and you may think this is an unfair question, but I want to ask why Pina has only made one film, *Lament of the Empress* (1989). Was that just a one-off venture? Do you have any ideas why?

Peter Pabst: Yes of course this is an unfair question, because you should ask Pina not me. I think she enjoyed doing it; but it is difficult to make a film. It takes a long time preparing, and a long time to shoot it. Then there's a long period editing it. I remember when I designed my last film, I was running mad because I was used to the quicker rhythm of theatre, where every eight or ten weeks you start a new production, and there I was working for nine months on the same production, dealing with the same issues. So film takes a long time, and Pina has a company of 30 dancers, and all in all 50 people to manage, so that doesn't leave much time available. And of course the running of the company is her first priority.

There is one thing that might be a reason for not doing a film. There is a contradiction between, for me, one of the greatest qualities of Pina's work, and the nature of film-making, and this quality, in my opinion, is her generosity. Let me put it like this. One doesn't often have good ideas. So if you have finally got one, you are very proud and tempted to say, 'look everyone, here is my idea. Here it is, have you all seen it?' Pina is generous in that sense of throwing ideas around, of wasting them, of throwing them away. Often you have ten different things going on on-stage at the same time, so it's inevitable that you won't be able to see everything. So what you don't see is to some extent wasted. However, you can see these pieces again and again, and discover new things each time. I find this even as her collaborator. Sometimes I ask her, when I see a performance some years later, how long have the dancers been doing this? She looks angrily at me and says 'since the beginning'. I think this kind of generosity is very important. In film, however, you always have to edit, you always have to make choices. In her pieces the public make their own choices about what they see. It's what they look at, and how they put things together. In film, I think you no longer allow the spectator that freedom, those choices.

Valerie: Returning to what you said about the salt, and the problems for the dancers, what's your relationship with them when you bring in difficult materials that they have to deal with it? How much give and take is there?

Peter Pabst: Yes, it is complicated. They have all the admiration I could ever give them, because I think they are really strong. One has to remember that they rehearse and they are dancers. They work like mad for months on a plain lino dance floor, and then I come, all of a sudden – and usually I come in fairly late – and introduce things. For example, in *Palermo, Palermo* there is the wall crashing down, weighing five tons and creating an earthquake when it falls down, because it's a real brick wall. Then, all of a sudden, the stage is covered with bricks, and they have that to perform on. It's very difficult for them. Or, all of a sudden they have salt, or they have sand or whatever. Rolf used difficult materials. The dancers go very, very far with what they accept.

I remember we once did a production, *Ein Trauerspeil* (*Mourning Play*, 1994) and I had the idea of a moving floor for the dancers to perform on. So I created an island, approximately ten meters by ten meters which was floating on water, really floating, moving around constantly with the movements of the dancers who were dancing on it. Because I wanted it to look nice, I got some material that looked like sand, but is slag from a blast furnace. It is a bit like ashes but very hard, and is passed through special mills. It has a consistency like broken glass, but is very dark, it's like black sand that glitters, shines and reflects. It's extremely beautiful. But the dancers had cut knees, cut feet, scratched arms, cut hands and they were just bleeding all over. It was a disaster. At the end of the day, they protested and said it was impossible. We had to call in a doctor in order to find out. So I said 'OK, I apologize'. It was very strange, because immediately they felt ashamed and said 'Well no, wait a minute'. So I got a finer gradation of it. The finer it is, the less it hurts, although it was not as black and beautiful as the original material. So we had a compromise. For each performance, they used something like two miles of medical tape; they taped themselves everywhere including their feet and they accepted that.

I once had to give up; it sounds rather like a joke. I wanted to have a totally shiny dance floor, and they said 'No it's not possible'. They could not move well enough, and we could not use it. It is always a question of trust. We have worked for such a long time so closely together. They know I wouldn't use anything dangerous. For example, they know that if I use some material I try it first. Before putting salt on stage, I would eat it in order to know what it does to me, because if they fall into it or jump into it, they would breathe it in or get it in their mouths. It is the same if I use earth or ashes. In *Palmero, Palmero* for example, I wanted to have a rain of red sand that came from Sicily come down from the fly tower. I used the kind of red ashes they use in sports arenas. Every time I try it out, I have a physical or chemical analysis in order to make sure. If I want to use earth, I only take it if I know where it comes from, and that there is no chemical contamination. The dancers have to be able to trust me in these practical things, and know that I would not confront them with anything that could harm their health. They, of course, trust Pina, too, that she would never accept anything that would damage their art. So this kind of trust, I think, is the reason that they accept so much.

Another example of trust concerns the set for the piece *Vollmond* (*Full Moon*, 2006). This has a huge river about four meters wide which flows from stage left to stage right. At first you can hardly see it because it is very silent. The only way to go from up-stage to down-stage is across a huge rock that lies across the river. This is so big that you have to climb over it. In the last 30 minutes of the piece, they start to use the water and then there is rain that is as heavy as a monsoon. I don't think there has ever been so much rain on stage. The dancers are unbelievably wild as they dance in the river and out of it. They dance with such energy it's really spectacular. The real miracle of this design was their trust. They started to develop these violent and energetic dances when there was no water, and I had no idea how to do it. I had to find a dance floor that they could dance on when it was dry but which they could also dance on when it was wet. They did all this work when I didn't have any idea how to do it, trusting that I would be able to find a way.

Ramsay: Many of the sets that you design create a sensation for the performers – water on stage, or walking through carnations. In *Masurca Fogo* they build a hut on stage towards the end, and they all go into the hut and dance, and there's a sense of claustrophobia. We, in the audience, can only see through a small window at the front, that inside it's packed with 22 dancers. In this case, the design creates an experience for them, whereas there are other designs which create a visual experience for the audience, but not for the dancers, such as the projections. There is a wonderful image at the end of *Masurca Fogo*, with the projection of the waves breaking in the water. From the auditorium, the dancers appear to be immersed in water, but they themselves can't see it.

Peter Pabst: Yes, of course, I think there is an unusual sensation for the dancers. But there is something different happening. I was just thinking about what that means. Regarding the projection, they don't have the impression you have as the audience. They don't see the images, but they are affected by them nevertheless. Don't forget, all the floor is moving, so these projections go on to the floor and on the ceiling, as well as on the background and on to the left and right sides. What they have when they dance, is a constantly moving floor with changing images and changing light. It's a very strong sensation for them. It does affect them. But I don't think I should affect them all the time, as that would drive them mad.

Ramsay: It can occasionally be very difficult to keep watching the dancers because they're dwarfed sometimes by the projections. In *Danzon*

(1995), still photographs are projected during most of the piece but then towards the end Pina comes on to do a solo. There's a still projection of a fish and then magically it starts to move because it is not a photo at all but a film (see Figure 12). This huge fish is followed by a succession of other tropical fish which almost appear to kiss her head. It's almost difficult to watch her because one's so distracted by everything else.

Peter Pabst: Yes, this is the conundrum. This is why you can see it twice again, you can watch the film and you can watch her, but you know it doesn't happen by chance. We always try to keep a balance of not losing a dancer entirely. You always lose something and you gain

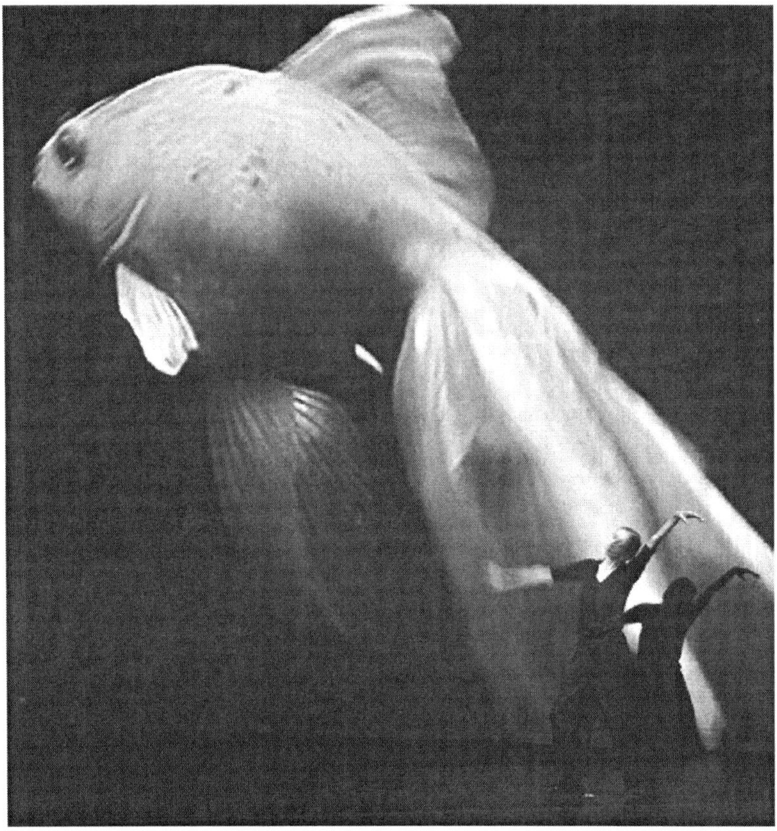

Figure 12 *Danzon* Choreography: Pina Bausch, Tanztheater Wuppertal, Dancer: Pina Bausch

Photo: Jochen Viehoff.

something. This is the story about the set for Pina's solo. This was the very first time I used video on stage and I remember I had very little time, because I came from another production and joined them just four or five weeks before the premiere, and Pina greeted me by telling me she had thought of going on stage again. I thought this is very bad, because now all the world will look at her just in this production, where I've got only four weeks to collaborate and what am I going to do now? She told me it was probably about fish. I thought well this is rather easily done, I've got to go to some video library and get a nice submarine film with beautiful fish and rent it out, and project it behind her. But there was nothing suitable. Such films are made for schools or for TV, and there is a commentary and camera men are visible, and things like that. It gradually dawned on me that I might be obliged to make this film myself. I only realised this about two weeks before the premiere.

One thinks a designer has to design and to control everything, but that's impossible. You always discover through doing, and find things you had never imagined. In this film, I found out first of all it must be nice fish and nice images. By chance I followed this fish with the camera a tiny bit, as it was moving up and down. When this was projected large behind Pina, it made her fly. It was as if she was standing in an elevator, flying up and down, and flying through the space. She stood with her feet and legs still and only moved her arms and upper body. Because of the video projection, she appeared to be flying around. You can't invent these things. Sometimes they happen, and you have to be open to see them.

The little hut in *Masurca Fogo* is another example. We had been invited to a party in a favella in the Cape Verde Islands, where these very poor people live. They have a law in Portugal where they are not allowed to build anything new. So the only way to build a house, or to add a room to an existing one, is to build it overnight, so no-one sees it. There is a gap in their law which says that what's done is done. This was where the idea of the hut came from. I was thinking about how to design the poor people's house, but this is already a contradiction in itself, since they don't design it. Pina wanted the dancers to build a house within five minutes. Fortunately, without giving me any chance for preparation, suddenly in one afternoon's rehearsal, she asked me if we could try this out. But I had not designed anything, because I still didn't know how to design a poor man's house. So I had about three minutes to chase down into the cellar where they keep all the rubbish and get something out that might be useful for building a house. This is what you see in the production. It would never have been so beautiful if I had had the opportunity to design it. It would have been totally boring. It is just so much more alive because it happened like that. This is another example about the limits of conventional design. One cannot and one must not be totally controlling all the time.

Ramsay: Many of your designs use landscape imagery. This is already present in *1980* with the grass and *Nelken* with its flowers, but the late work has much more explicit landscape elements such as the rocks in *Masurca Fogo*, the waterfall in *Ein Trauerspeil* and the river in *Vollmond*. Could you say something about this?

Peter Pabst: In the end the imagery is never realistic. It is materialistic in that we use real materials. There is a never ending tension between nature and this artificial space that is called theatre. Nature is something that can never be consumed. It is like if you compare wood with steel. You can take a piece of wood and you can bend it as many times as you want, as long as you don't break it. If you do the same with steel, it will last something like five minutes and then it will break. It is like the music in the opera. You have to go to endless rehearsals and run throughs, but the music is like nature: every time you come with joy because of the music you are going to hear. Nature can never be consumed.

Ramsay: At the end of *Palermo, Palermo* there are upside-down trees that are lowered from the wings. For me there is something poignant about these trees because they have been cut. It seems an image that arouses ecological concern.

Peter Pabst: This is something that you can say but I can't. But for me it is a kind of poetic image because on the one hand they are upside down but on the other hand they are flowering. They're covered in cherry blossom. It is a contradictory image. There are these beautiful pink flowers but nevertheless they are upside down.

Valerie: I think we should open the questions to the audience.

Audience: You said that you always start without an idea, without anything. It's very hard when you're trying to get space or money, or just people to collaborate, they always want an exact idea. How did Pina start off?

Peter Pabst: I do not really know how she started out, probably not exactly as I've described it, because I joined the company in 1980. The company was founded in 1973, so seven years earlier. I don't think she had developed a system. But there is one big privilege in the system of subsidised German theatres. Where a company is engaged and paid for, the artists have long-term contracts, they don't have to look for money, they don't have to present a project to sponsors. It is hard to think about what it would be like if they had to do that, which is normal in other countries.

Audience: Concerning your role during the devising process, does your response in terms of design come from what the dancers come up with in improvisation? Are you an active part of that? For example, as part of an improvisation, do you bring in some soil, or some salt for them to work with? Does Pina Bausch aim questions at you as well as the performers?

Peter Pabst: No I don't do that. It influences me, sometimes I help with something, or I try something out. But that's different, it's by watching, and then I think sometimes they invent it entirely. I don't have to invent; I just have to take care that they have everything they need and that they can use it. Sometimes they try something, and I see. It's like a ping-pong game. The ball comes over here, you try to catch it. Then you offer something and they take it and they do something probably unexpected for you, so the ball comes back. This is very interesting. To give you an example, we have done one piece called *Ahnen* (1987), which means 'Ancestors' but there is a double meaning in German, making the title more complicated. The set is a huge wood of cactuses six, seven meters high as if in the Arizona desert. One day I saw one of the dancers in the rehearsal studio playing around with a little tele-commando, remote control car. I watched her playing and I thought, 'that's nice'. I don't know why but it made me think about a remote control helicopter. I did a little sketch and showed Pina, and she said 'yes, that's nice'. I didn't realise what I'd done, and what a mess I was in, because then I found that to fly a remote control helicopter is extremely complicated. It takes about two or three years in order to be able to safely get it to take off and land. Now we wanted to have it fly around this wood of cactus. They are huge, they are dangerous, and if the helicopter were to escape, it could just cut off your head.

By chance I found out that, about half a mile away from the Wuppertal Theatre, there was a little shop that sold model ships, and airplane and car models. Its owner, a huge man, was the German champion of remote control helicopter flying. So I simply had to sit at his feet for about three days and convince him that he had to get up (which was difficult enough) and bring his helicopter and come with me to the theatre. This is an example of that kind of ping-pong thing.

Pina will look at a model of a set and ask me 'and what else can it do?' and then I promise to bring down the blue out of the sky for her. In the Hungarian piece *Wiesenland* I made something extremely beautiful. It is a huge wall completely covered with green plants and moss and other natural materials; water pours out of it like little chains of pearls. The falling water makes a variety of little sounds which depend on whether it lands on a pool of water or on to a rock or on to a plant's leaves. It is huge, about 40 meters wide and ten high. When I showed it to Pina she asked 'what else can it do?' So I said to her we could put it down on the floor so that they could dance on it. And she said yes but what else? So I said we could put it up so that they could dance in the drips from it. Luckily she was not interested in putting it up, as working out how to lower it down became a nightmare. I thought I'd found something that was very simple to make and easy to handle so we could do it easily. But three technical directors ran away and I engaged someone to resolve the technical problems and after a day he disappeared. In the end I found him lying down on his hotel bed in a dark room. He said it is going to fall down. I had thought it would be too easy if it just moved straight down and that it would be more interesting if it turned as it was lowered so that it was no longer parallel to the front of the stage. It turned out that it weighed between five and a half, and six tons. Everybody said no. They sent the city government to hang round my neck, and the insurance company, and the inspectors of health and safety in the work place. But in the end we sorted out the problems and it is very beautiful. I was just happy in the end that Pina didn't want it to hang upside down as well. It is just one example where Pina asks 'what can they do with it?' and I promise something impossible and then wonder how am I going to do that.

Audience: What do you have to consider when you choose whether to use a video or a light to show something on stage? What do you consider when making a visual choice?

Peter Pabst: It's rather difficult. It's not such a clear choice. Obviously a blue line doesn't do the same thing or create the same kind of image in a stage space as something else would. Nevertheless there is a choice. The problem is that it happens when the piece is not yet ready. It's all try outs, and it could help, or it could create something to add, in this case, a video projection. Sometimes this is so complicated that you can't do it, because you might hide everything we have spoken about. Sometimes it's difficult to see what the performers are doing if so many other things are happening. It would be too complicated. You would destroy everything by adding a film or a video. Sometimes it needs something very silent, very subtle and very smooth but at other times it needs excitement. It's more about these atmospheres, than making an intellectual or conceptual decision.

Audience: Is there anything that you would really like to do but haven't had the opportunity to, yet, in design, in crafting space?

Peter Pabst: There are things I've been dreaming about for a long time, and I still don't know how to do them. For a long time I wanted to have a waterfall and I didn't know how to do it, for the practical reason that water is very heavy, and for a strong waterfall, you need a lot of water falling down. I knew I couldn't put two tons, or three tons or ten tons of water on to the gallery, it would just come down immediately, including the gallery. These ideas stayed with me for years and years until I did this thing with the floating island. I had a little problem: it was strictly forbidden to use one drop of water on that stage and I had calculated, I needed about 23 thousand litres in order to make the island float. I thought the only way of resolving the problem was to see the fire brigade. They were responsible, because there is electricity underneath. I made an appointment with them and prepared some nice little models for them. I showed these to the head of the safety department and said I have no idea how to do this. I think there is only one man who knows, and that is you. So I'm in your hands. They fought for that, it became a very funny story. It's very tough if you know you have two weeks to realise all of that against many opposing forces. I didn't know what a floating island does and when it floats, I could think about it and I could hope that I had thought properly enough, but I knew if it didn't work I would only know about it three days before the premiere. This is tough. So they helped me and we did it. I had an obligation to empty all of the water out of the theatre within, I don't know, five minutes. The different safety departments involved thought they had killed it by imposing this requirement. The fireman, however, gave me a huge pump which could pump 23 tons of water outside the theatre in five minutes. So I thought if it can do that then it should be able to pump the same amount of water on stage and have it fall down, and this was my waterfall! So sometimes you have to carry a dream for a long time. I'm dreaming of having a set that starts to burn in the beginning and it will be burnt at the end and I can use it again the next day, but I don't know how to do it yet.

9
Interview with Meredith Monk about *Turtle Dreams (Cabaret)**

Valerie A. Briginshaw and Ramsay Burt

Introduction

'Enjoy yourselves, folks!' Pablo Vela is the MC for Meredith Monk's *Turtle Dreams* (*Cabaret*, 1984). He stands at a mike on one side of the stage while a couple of electric keyboards stand on the other. In the space behind them are two red plastic quoits like roadwork cones, lit from the inside. The scene is set for a cabaret. But the opening piece, the sinister *Engine Steps* couldn't have been less cabaret-like. Dressed all in black with an outsized circular hat like a leaden halo, Gail Turner stands close to the front row of the audience, her hands lit by a tight spotlight. She joins her thumbs and forefingers to make a diamond shape, which she squeezes and shakes. Every now and then she rotates her hands to form a transparent globe, or wiggles all the fingers loose like a sea urchin. Over the sound system comes the noise of clanking machinery – a railway locomotive? These two scenes give the two sides of *Turtle Dreams (Cabaret)*. On the one hand, the convivial informality of a series of cabaret acts in costumes with sequins and satin facings. On the other, chilly intense rituals. It is the songs that hold the evening together, sung directly to the audience. All the rest builds up from them. Coming as part of the Dance Umbrella, some people were apparently surprised there was so little dancing. But *Turtle Dreams* is full of familiar Monk themes.

All the opening pieces in the cabaret are quite short – the opening sequence, *Two Men Walking*, *Engine Steps II* and two short films. But closing the first half and opening the second there are two longer pieces, the *Tokyo Cha Cha* and the *Berlin Waltz*. *Tokyo Cha Cha* starts with the

*From *New Dance*, No. 39, New Year 1987, pp. 11–14

three men making mouth percussion noises into microphones, who are joined on the other side by three women forming another backing group. All the performers wear green or blue loose jackets with satin collars and matching trousers, and each has a little black cone with a red light at its tip, sticking out from somewhere – a shoulder, the back of the head, an arm. The piece builds up slowly. The song is sprightly and funny with inane words 'Let's cha cha, all happy, all beautiful day, you happy, me happy' and the performers sway as they sing round the microphones. Gradually they spread out into the space, forming lines. They shuffle and shunt with little steps and develop a sort of fluid arm jive routine. The lines form and reform vertically and horizontally; the performers grin 'geisha' grins, bow to one another, take their hands out of their pockets and sprinkle black confetti. The throwing looks carefree, but as it falls the confetti looks ominous, like charred flakes falling after the bomb has dropped. Pablo Vela fills a line of glasses behind the singer – the water from his jug magically turns a different colour in each glass. Later he walks across the back carrying a four-foot long white model aeroplane. Towards the end, the performers bop merrily across the stage and blow kisses to the audience, then sprinkle more black confetti.

After the interval, the *Berlin Waltz* has a driving insistent rhythm and is more wailed than sung. Most of the words, as usual with Monk, are indistinct, but occasionally one picks out a few 'I went to the store', 'Fire, fire, fire, fire'. Four dancers; two male, two female, in smart black suits and dresses, stand in a chevron formation. They all step to the left, then tack to the right, to the left, to the right, in a waltz rhythm in stark unison, staring straight at the audience. The choreography consists of very formal unison, linear stepping patterns with angular movements of the arms and upper bodies. When the music momentarily stops, the shoes carry on beating out the rhythm. Sometimes they group into mixed couples, sometimes the women stop and look at the men, but it is unrelieved, bleak and frontal. When the performers turn sideways, we catch a glimpse of a diagonal shower of silver studs sewn, like on motorcycle jackets, to the backs of their dresses or jackets. One figure turns to face the back with arms outstretched in a Christ-like pose, then joins back into the rhythm again. The repetitive, cyclical form of the piece, like the circular structures of Monk's music, becomes quite obsessive. Finally, into this tense atmosphere, waltzes Gail Turner, all in white with silver shoes, her dress stiffened with wire into a cone shape. Her arms up, her face dreamy, she dances a flowing circular waltz, gazing into the eyes of a non-existent partner. Behind her now at the back, like a backing group, the four stamp sideways bending over as if ducking out of the way of something, still stuck in their deadly mechanical waltz.

There is little after this. Another black and white film, this time of a deserted city. After a few shots of empty streets, a giant turtle potters round a corner and you realise the streets are miniature models. Monk comes on wearing trousers at least ten foot long, carried behind her, like a bride's veil, by Pablo Vela. She sits on the floor with a slide of the moon behind her and sings a lullaby to end the show.

While she was in London, Meredith Monk very kindly agreed to talk to us about *Turtle Dreams (Cabaret).*

Q: At the start of *Turtle Dreams (Cabaret),* I was particularly interested in the opening moves that Gail (Turner) does – for me that was definitely a women's symbol.

Meredith Monk: People have said that. My father said that, but that was not where that came from. It's interesting. Where it came from was that we were working on demon characters. That was her demon character, but in the end it turned out to be more of a conjurer or magician image. I was thinking of it more like a crystal ball or centre of the body. Then there were the two men who were in green clothes with their heads covered, also doing frenetic kind of movement (*Engine Steps No 2*). That was also working on demon characters, but what that ended up with was more like men that are electrocuted with electric sparks coming out of them. We were working with that idea of possession where you do movement where you really go out of control.

Q: But there seem to be male/female references throughout the piece. Circles against lines: the lullaby at the end in front of the moon, is female and nurturing. The two men walking across the back earlier are very angular and threatening.

MM: I wasn't conscious of that when I was working on the piece. I was conscious of a happy/sad thing: a poignant thing about the world. *Turtle Dreams* is strangely enough a little like an elegy. How can you do a piece that is so much fun but has such a sadness to it? For the cabaret feeling it really needs to have that buoyant humour and fast energy. This piece has pretty fast energy compared to my other theatre pieces which are much quieter. The *Tokyo Cha-Cha* is really the most extreme with hilarity; an hysterical kind of humour, and I hope very witty in the way it's put together, but it has a real scary thing underneath. I wasn't really conscious of a male/female thing, but I did notice while I was working how the circle at the beginning became the circular screen and ended up being the circular moon. I like to work that way, so that things repeat but in variations. I always think of those as visual rhymes.

Q: But isn't your work critical of the present situation, with what we have to put up with?

MM: Yes definitely. I work in a pretty oblique way, though. I don't think my skills lie in being a political analyst in the way that Brecht was, because I'm really a musician. I come from a musical background, and even if I am working in images, it's musical. You could say it's a kind of poetry – working in poetic forms rather than prose. I don't really like to use that much text, I don't do that kind of analysis. It's more getting into a feeling level. But there is always some kind of political layer in everything I do. It is one layer, though, one strand woven through, I think, in any piece I've done. *Vessel* (1971) had a Joan of Arc theme, but it was also very much about contemporary life, it was dealing with motorcycles and technology as well. It was almost as if you were looking at this time from a medieval point of view. In *Girlchild* (*Education of the Girlchild*, 1973) – people have asked whether that was a sort of feminist piece, and I would say it was a kind of humanist piece in that each of the seven women was such a remarkable character that it showed women human beings in a very positive way. It was not an angry statement. It is just that if you saw those seven women sitting together it was powerful and rich. So I think that there is a layer of political consciousness but it's not the main thrust of my work.

Q: What about all the references to high technology and to the urban environment in *Turtle Dreams*?

MM: I always feel that an artist is trying to be, in a way, connected to something that is the undercurrent of the unconscious, or what's going on in our culture: a bit like being an antenna so that you're picking up essences of what is going on. I don't think that I want to do art that is a mirror of society, at all. I am interested in essentialising those currents and then doing one thing or another with them. One thing is, I think *Turtle Dreams (Cabaret)* is like a warning piece, and some of the other things we have done have been that in the last few years, like *The Games* (1984) and *Specimen Days* (1981) and *Recent Ruins* (1979). A piece that I have just finished in New York, a piece called *Acts from Under and Above* (1986), is almost the opposite. It is closer to *Education of the Girlchild* and some of the other pieces, which offer a human alternative. But it took me, going through all those recent warning pieces, to be able to go back to a new synthesis of what I think an alternative is, without it being what we had been fed in the sixties and seventies, of what an alternative is. As far as I'm concerned, the only thing that *Acts from Under and Above* is, is simply a piece about a long, long friendship with Larry Harrison who I have known and worked with for many years. Very simple things, and a very strong sense of intimacy.

Q: When you talked about the performance on Thursday night to the audience, you said that you wanted the *Tokyo Cha-Cha* to be an entertainment, and that on some levels you just wanted people to sit back and enjoy it.

MM: On some levels, but that's too simple. [laughs] I don't like to let people off the hook that easily! I like it when you see something and there are all sorts of levels going on simultaneously. So I've tried to compress as much as I can so that there are a lot of different perceptions going simultaneously.

Q: How did you get to using cabaret?

MM: How I got to it really was that around 1979 I became increasingly dissatisfied with theatre forms, even though I did another theatre piece after that; *Specimen Days*. I was doing a lot of straight music concerts with my ensemble at that time, and really appreciated not having to play one role the whole evening, and I liked being able to talk directly to the audience. Also, in a way, I was able to start with myself as my persona, and then start building different personas within each song, so that it became more flexible. It seemed at that point to be more honest to me as well. The level of meaning came not from a literary source, but much more because music is a kind of heart to heart communication, and people will either respond to it emotionally, or not. We were doing a lot of music performances like that. Then a few years later I began to think I wanted to try to add some of the other levels into this music concert format. So, I thought, what sort of theatrical format has a kind of unit by unit structure? and that's how I started to think about cabaret.

Q: You tend to use modal rather than diatonic scales in your music. Could you briefly explain the difference?

MM: Diatonic music refers to functional harmony, which is the basis of Western music. Modal music is based on scales that might include major or minor, but don't necessarily relate to a chord system. Medieval modes are represented by the white keys on a keyboard. Each mode has a slightly different character. A lot of early music is written modally including Greek music – that's how the modes were named, like the Phrygian mode. Modal scales usually have a more poignant quality to them. I do usually write modally, although I have written in a minor key diatonic scale, and one or two pieces in a major key.

Q: Why particularly do you use modal scales? Is it wanting to get away from Western technological civilisation to something before it and after it?

MM: Before it and after it, yes. [laughs] I think that that scale structure has a lot to do with trying to find a music that is cross-cultural, so that it includes Western culture, but is not really limited to Western cultural possibilities.

Q: They are also sung musical cultures, rather than instrumental ones?

MM: There are instruments in them, but otherwise, yes. I was reading a very interesting book called *The Timing of the World* by a Canadian composer, Murray Scheiffer, who is also a sound archivist. He's collecting sounds in an archive; sounds that are literally disappearing. For example, he's taping certain kinds of trains that are becoming obsolete. He talks a lot about the fact that Western music has become a visual music, which is its real limitation. In other words, Western music has become more about paper than hearing. I always say it in a different way. To me what Western classical music in the fifties was coming to was as far as it could go from the chin up. Music for the head, a sort of isolated head bobbing around with no body at all. I'm not interested in that at all. So in a way, different scales are a way of getting back to a more oral kind of music – more physical and more kinetic. To include the mind, of course, but to have the whole body, the heart, the mind, and the spirit in one.

Q: In the foyer at Riverside before the show, there are the two films being shown, and there is the woman at a light box moving marbles around on a map. What are you trying to do with them?

MM: I love the idea of creating a world when the audience walks in. In most of the pieces, which have a theatrical element, I try to have some kind of a pre-set, the idea that there is something going on, before you go in. The piece is happening all the time, and you've just walked in on this thing.

Q: What was the starting point for *Turtle Dreams (Cabaret)*?

MM: The *Berlin Waltz* was the first piece that I worked out. It was a very interesting piece to work on, because it did create its own rules and they couldn't be broken. The sensation was that the piece was already made and that we were just trying to find it, and wanted it to talk to us, and tell us what it was. Its structure is a very abstract, rigorous form, in which very intense emotional units hit against this structure all the time.

Q: When Gail, the woman in white, comes in, you are all moving on very straight lines and doing angular movements, and she comes in, in the hooped skirt, and takes the waltz actually round in curves [...]

MM: Yes, you finally get a real waltz. You've waited the whole piece for it! She is this fantasy waltz character, if you really want to get into some of this content (the way you can take a piece like this, but it certainly isn't the only direction you can go in). As soon as I start thinking about the Berlin aspect of the waltz, then I start thinking, well Gail is the perfect Aryan specimen waltzing in, blonde and pure and all that. She also looks like a *Breck* hair ad. from the fifties! [laughter] We always thought of her as an angel in white cutting through an emotional field. I think of the waltz as an emotional force field. It also has those images of [...] like the bomb has dropped and people are looking for one another in the fog. What I like about that is that you get into those sort of images but then it goes right back into a more abstract sort of thing. Little shadows, or snippets of images.

Q: The bomb-dropped images also come out in the little film that is shown on the hoop.

MM: You know what I was thinking of in that film? I was thinking of Pompeii too. Have you ever seen photographs of those figures caught in a position by the lava?

Q: That brings me to your fascination with a time scale – with a past and a future – in your work. It seems to come up in so many of your pieces.

MM: I don't know why. I really don't. It seems to be something I've always been concerned about. Time and space. Disjunctive time and space.

Q: Is it to do with what you said about creating a dream world?

MM: I think it comes down to offering something that brings you into a world that has its own integrity, so that, when you left it, you would be able to go back into this world and you would see it in a fresher way, a more vital way, a more attentive way. In terms of music, I think that I try as much as possible to offer a very full palette of emotions, which I think vocal music can do, that cannot be called into play all the time in daily life, but is there. This is one thing that art can do, which is to make you look at what you take for granted in different ways.

Afterword

This interview was first published in *New Dance* magazine in 1987. During the 1980s, we were both regular contributors to *New Dance* which acted as a seedbed for much of our early critical writing, as it did for many other dance writers at the time. Founded in 1976 by the X6 Dance Collective, it was the only publication where independent dancers could write about their own and colleagues' work, much of which was either being ignored or misinterpreted by mainstream critics. Its libertarian, alternative, feminist ethos promoted the exploration of new ideas and new approaches to dance writing. *New Dance* was produced collectively. This encouraged collaborative working methods which have been the foundation of our writing relationship over the years.

This interview is the oldest piece in the book. Why after more than 20 years did we think it worth re-publishing? As with some other artists whose work features here, we have both enthused and written about Monk's work since the early 1980s. Despite the interview's age, it is still most pertinent today. Not only does it complement Chapter 10, Ramsay's *'Danced Testimonies'*, which discusses Monk's *Ellis Island,* it also anticipates themes discussed throughout the book.

Radical experimental work is almost inevitably interdisciplinary which is why much of the contemporary work we discuss here blurs disciplinary boundaries. Monk has moved further in that direction than any of the other artists whose work we discuss. Her declared interest in disjunctive time and space, evident in her exploration of archaeology in *Ellis Island* discussed by Ramsay, and her account of making the *Berlin Waltz* in the interview, resonates with our themes of rethinking temporality and insurmountable memories. Monk's concern with multiple levels and perceptions resonates with Deleuzian theories which inform other pieces in the book. There is evidence throughout the interview of the importance for Monk of affective intensities, also discussed in Ramsay's upcoming chapter. These point to distinctly human values associated with emotions, friendship and intimacy, which in turn ethically inform her declared political stance, evident in her desire to create alternatives and her belief in art's power to help us see things differently. All these concerns also underpin this book.

Section 5
Insurmountable Memories

Introduction

The two chapters in this section exemplify another instance where we have each come from very different directions towards similar sets of concerns. Both chapters discuss works that, each in their different ways, address violence and insurmountable memories. In Ramsay's chapter, Meredith Monk's *Ellis Island* and Isaac Julien's *Western Union* rediscover the violence accompanying the processes of migration that has been largely excised from cultural memory in the United States and Europe. Valerie examines the way Emilyn Claid, through her process of making *Remember to Forget*, rediscovers a violent potential within dance movement that is invariably hidden or finessed away in much contemporary choreography. Ramsay therefore considers instances in which the dancing body functions as a site of cultural memory, while in Valerie's chapter, embodied memories are an integral part of the process of creating dance movement. In Claid's dance-making process, the memories involved may not be violent but their evocation is often a means through which movements are created that produce a violence of sensation.

These different kinds of memories, in both chapters, all invoke suppressed histories. In calling these memories insurmountable, we are not inferring that it is impossible to progress beyond the violence that they can entail. They are insurmountable because they are eternally there. As a consequence of their embodied nature, they keep coming back to haunt us. They won't go away. We are not saying that any of the pieces we discuss in this section are *about* memory as such. What we are pointing to is the important roles that memories play in the sensations and perceptions involved in experiencing some dancing. The pieces Ramsay discusses in Chapter 10 seem to be about the socio-political power of these insurmountable memories, but what he demonstrates is the extent to which their ability to move us depends upon affect. *Remember to Forget*, however, seems to be about the affects that the violence of sensation brings forth on a personal and interpersonal level. Valerie suggests in Chapter 11 that discovering the complexity and plurality of these sensations and affects, often contaminated by memories, can have political implications.

10
Danced Testimonies of the Traumas of Migration[1]

Ramsay Burt

This chapter examines the way two interdisciplinary works that combine film and dance – Meredith Monk's film *Ellis Island* (1981) and Isaac Julien's installation *Western Union: Small Boats* (2007) – explore the lingering after-effects of the experience of migration. *Ellis Island* was filmed in the former immigrant reception centre in New York harbour and takes as its premise a spectral archaeology. It asks: what ghostly traces of the migrant origins of the families of many of today's US citizens might such an archaeology uncover? *Western Union: Small Boats* is a multi-screen video installation occupying three rooms, made in collaboration with choreographer Russell Maliphant. Filmed at locations in Sicily, its subject is the plight of clandestine sub-Saharan African migrants, thousands of whom sail in small boats across the Mediterranean in desperate attempts to enter the European Union. Both works were closely connected with live theatrical performances during which some of the film footage was projected. US citizens, with the exception of Native Americans, are all descended from migrants, many of whom entered the country through Ellis Island. Sicily has a long history of migration including Phoenician, Carthaginian, Greek and Islamic settlers who, over the centuries, ruled Sicily. Migration today is often represented as a painful and disorienting experience that needs to be gone through in order to reap the promised future rewards of integration within the new, more modern, host country. Monk's and Julien's films draw on the tradition of US modern dance in ways that complicate this modernist narrative of progress and assimilation. While not suggesting that they are against either modernism or progress, in my view, these films use sites of memory to propose alternative ways of looking at migration. These, I shall show, correspond in significant ways with an account proposed by the philosopher Vilém Flusser of the homelessness experienced by the migrant.

Flusser left his native Czechoslovakia in 1939 to escape the holocaust, and then, after 30 years living in Brazil, migrated again due to unease at living under a military dictatorship. For the rest of his life, he lived in the south of France. He was, therefore, someone who migrated from the Old to the New World and back again. This experience led him to argue that migration is liberating because it frees individuals from the unconscious threads that tie them to a homeland and national identity. His family, who had remained in Prague, died in concentration camps. His Czech friends all died in the Resistance, and all his German Czech friends died on the Russian front. Prague, he wrote, had nothing left for him, but he had found Sao Paulo much more difficult to leave. What the migrant experiences, he suggested, is not 'cutting off all relationships with others, but weaving these connections in co-operation with them. The migrant does not become free by denying his lost home but by overcoming it' (Flusser 2002: 95). Flusser's surprisingly positive view of the migrant's state of homelessness is one that is framed in ways that are explicitly critical of the modernist ideologies underpinning nationalism. It is this critical attitude that, I suggest, underlies the way these two works by Julien and Monk approach their subjects.

The descendents of migrants are shown, in both works, acknowledging rather than suppressing or denying memories of lost homes. These are indicated in two different ways. They are brought into play because of the associations of particular places: Ellis Island in Monk's film, the coast of Sicily and the Palazzo Valguarnera-Gangi in Palermo in Julien's installation. They are also evoked through dancing. To embrace homelessness not only means freeing oneself from the national ties of the past; in Flusser's view it also creates the possibility of escaping any further ones. The migrant, he argued:

> is a man of the coming future world without homes. In his subconsciousness, he carries the mysteries of all the homes he has once passed through. [...] he gives evidence to the native that there are not only numerous homes, but also numerous mysteries. Moreover, he is evidence that there will not be any more mysteries of this sort in the near future.
>
> (Flusser 2002: 102)

These mysteries are a consequence of being settled and identifying as a member of a particular nation. The migrant, who does not feel any such ties of identity, is thus without mystery: 'He lives in the clarity of the fact of being, not in mysteries. He is both a window through which those who have been left behind may see the world and the mirror in which they may see themselves, even if in distortion' (Flusser 2003: 14). My aim in this chapter is to examine the way these two works explore the kinds of relationships between migrants and settlers that Flusser has discussed. I will suggest that the works themselves function as the kinds of windows and mirrors that Flusser identifies and, by doing so, remind their audiences of their own potential for relating, in ethical ways, with others.

Monk's 30-minute film *Ellis Island* was developed from seven minutes of filmed material originally shown during her stage performance *Recent Ruins* (1979). Ellis Island in New York harbour was a centre through which, between 1892 and 1927, almost 16 million migrants arrived hoping to be allowed entry to the United States. Abandoned in 1954, it is now a museum but was still a ruin around which occasional tours were led when Monk was making her film. When the original, short version of the film was shown during *Recent Ruins*, a group of performers who played the role of migrants in the film, passed through the theatre space dressed as present day tourists on a tour round the centre. In this piece, Ellis Island was one of the ruined sites in which archaeologists laboured to assemble meaningless fragments into evidence of a lost past. In one scene, an older archaeologist (Pablo Vela) teaches a younger one (Monk herself) to reconstruct a vase from shards while on a screen behind them a disembodied hand uses a marker pen to draw the outlines of ancient and modern household artefacts (see Maranca 1980). Monk's theatrical presentation of archaeologists seems almost to anticipate Mike Pearson and Michael Shanks's later theorisation of the performative nature of archaeology as a discipline. A site specific performance, they suggest, recontextualises its location: 'it is the latest occupation of a location at which other occupations – their material traces and histories – are still apparent' (2001: 23). The film, I suggest, records a site specific performance whose affective power derives from this kind of recontextualisation.

The final version of the film weaves together scenes, filmed in colour, of imaginary present-day tours with creative re-enactments, shot in black and white, of past scenes of migrants. The sound track includes unaccompanied, wordless vocal music composed by Monk and sung by her vocal ensemble. In a loosely narrative way, the film alternates between the present and flashbacks to the past which become complicated when figures from the present begin appearing in them. The migrants seem to inhabit a world that has lost its centre and in which the meanings of things have been transposed or subjected, on a number of levels, to confusing substitutions. In this vulnerable state, they are subjected to institutionalized, 'scientific' examinations that are medical, ethnographic or eugenic. These are carried out both by officials from the time when Ellis Island was still operating as a reception centre, and by time-travelling archaeologists from 1981, whose striped measuring rods appear in many of the early shots in the film.

In an early scene, the camera makes a wide panoramic shot from one corner of a dusty, ruined shed to another discovering along the way a group of migrants posing warily for a formal photograph. Concerned that they should look their best, one woman picks a speck of lint from the lapel of her husband's overcoat. Another shot shows a fretful old woman with two girls posing for the photographer with large, numbered labels attached to their shawls. Posed formal photographs are followed by views of food set on a plain, white plate with a rather unhygienic-looking old spoon: beans stand for a Mexican couple, spaghetti for an Italian youth, potatoes for an Irish family.

In another sequence, a large, white medical ward is initially shown empty. Then the image of a woman whose eye will be inspected appears as a transparent trace that slowly solidifies. Other migrants and medical personnel walk into the room and freeze in positions as if in the middle of some actions. Then, finally, they all come to life. Standing half-hidden behind the woman is a woman in modern black trousers and T-shirt who is her 'shadow', bending back and lifting her head when the migrant woman bends for the doctor to inspect the health of her eyeball. Another shadowy figure copies the movements of another migrant in a half-curtained cubicle in the background. These shadowy figures inside the scenes are 'archaeologists' from the film's 'present'. The camera's surveillance retraces the patterns of institutionalized dominance enacted by the officials who originally inspected the migrants. An impassive male warden with folded arms looms over an old woman rearranging four pieces of card at a small table in what appears to be an intelligence test. All permutations seem equally felicitous to her, whereas to a 'modern' Western eye it is obvious that the 'right' arrangement of pieces will make a white square on a black background. Modernity is a mystery to the old woman, and for other early twentieth-century migrants shown later in a classroom trying to learn the names of late twentieth century, suburban, consumer goods like microwaves and snippers. In Flusser's terms, the film presents a mirror in which settled, modern citizens can see themselves from an unfamiliar point of view.

Like the archaeologists, the film itself thus tries unsuccessfully to reconstruct and understand the ghostly fragments of the past history of Ellis Island, just as the officials in the past tried unsuccessfully to investigate and classify the migrants. The doctors, warders and teachers in the film who represent 'modern' Western values seem to approach their charges using methodologies that probably inspire uneasy feelings in twenty-first-century viewers. This uneasiness becomes explicit when a disembodied hand holding a marker pen interposes itself between the film camera and a migrant's head. It is as if someone is drawing on a photograph in order to analyse racial features, making measurements, and drawing attention to a nose or the angle between eye and ear. The hand writes on one face 'SERB', and on another 'J' for Jew (see Briginshaw 2001: 121). But little wavering movements of the head and blinks of an eyelid betray that this is not in fact a photograph, and that the migrant subject is actually there. The marks are being made on a sheet of glass. Is it an official in the past or an archaeologist from 1981 who is seeking out and marking these signs of racial difference and for what purposes? The more the film shows the dehumanising, modernist processes to which the migrants were subjected, the more it makes the film's beholders aware of the basic, common elements of humanity through which life survives.

Just as the film *Ellis Island* grew out of a stage performance *Recent Ruins*, so Isaac Julien's video installation *Western Union: Small Boats* (henceforth *Western Union*) was developed alongside *Small Boats*, the last part of an evening-length programme *Cast No Shadow* (2007), made in collaboration between Julien and the Russell Maliphant Company, that combined live dance with film.[2] The multi-screen installation, *Western Union*, which I saw at Metropictures in New York in November 2007, is a complex piece that includes all the footage shown in the theatre piece together with additional material.[3] Both juxtapose stunningly beautiful film of different kinds of events staged in Sicilian locations that suggest a complex and affecting meditation on issues concerning economic migration. As a theatrical performance, *Cast No Shadow* was an interesting experiment which generated intriguing new possibilities for combining live dance with film. At the time of writing, Julien and Maliphant hope to explore these further. The installation was, in my judgement, a richer and more assured work, and it is therefore on this that I am focusing my discussion.

Western Union is a multi-screen installation that occupies three galleries. The two outer galleries each contain photographs together with a single screen showing a relatively straightforward film. These flank a more complex, three-screen installation in the middle gallery which runs in an 18-minute loop. The curator Mark Nash suggests that the installation as a whole 'occurs on the borderline between fiction, documentary, reality and fantasy' (2008). While concurring with this, I would add three other key aspects of the work: the unusual presence of Rebecca Myrie in the film and photographs, the centrality of dancing throughout the work, and Julien's vivid use of colour. All of these, I argue, give an increased intensity to the experience of beholding the installation. Nash argues that Julien's approach 'does not abrogate an aura of political agency, although of course the work has a political dimension' (ibid.). In Europe, the topic of migration is an emotive one: right-wing politicians generate anxiety about the supposed threat that 'floods' of migrants pose to national identities, while on the left there are concerns that economic migrants, like those trying to enter Sicily, are victims of the effects of policy decisions made by Western governments and international institutions. The emotions aroused are affects that have been identified and fixed within a particular social or political context, whereas affects are more open and allusive. My proposal is that the intensity which Myrie, the dancers, and the use of colour bring to Julien's reflections on migration, keeps the affective meanings that the installation generates from becoming fixed to a particular political agenda, and opens them up to include wider questions about the ethics of relations between self and other.

Rebecca Myrie is an untrained performer who has appeared in a number of Julien's recent films and installations. *Western Union* is the last of a trilogy of installations; Myrie appeared in both of the previous parts of this, *True North* (2004) and *Fantôme Créole* (2005). In *Western Union* she is shown surveying the coast and beaches, and the Palazzo Valguarnera-Gangi, often seeming to meditate on locations previously occupied by migrants or to revisit scenes in the wake of the dancers. The way she is framed by the camera suggests that she could function, in Flusser's terms, as 'a window through which those who have been left behind may see the world' (2003: 14). As herself someone of African origin, she seems to witness scenes on behalf of others. This is a role that recurs in Julien's installations and films over the last twenty years.

An aspect of performers' presence (on stage or film) is their awareness of being looked at. This means that their awareness of themselves is doubled, complicating their self-relation. They not only do something but are aware of how they appear to a beholder as they do it. Professional dancers and actors are able to direct their attention towards what they are doing by focusing on whatever technical approach to performance they have learnt. This generally gives a certain distance to their presence. Untrained performers like Myrie have no technique to draw on when they experience this doubling of their self-relation. The more they perform, however, the more they learn to deal with this and thus lose a certain immediacy. The immediacy of Myrie's presence, as she has been filmed on the locations in Sicily, and the lack of distance in her self-relation gives a particular intensity to her performance.

The fluid, unfixable dance sequences in the installation generate qualities that range from the dynamic to the sensual. Julien told Martina Kudláček that what interests him about dancing is its potential for generating images that cannot be translated into words (2007: 79). Unlike verbal language, dance does not convey fixed meanings because it is fluid, always in transition, or indeed in transition between transitions. This fluidity allows choreography to create associations and correlations between imagery in open-ended and allusive ways. Dancing is not just one of the elements out of which *Western Union* is assembled, it is central to the dancerly way that associations and correlations are woven between different ideas and images presented in the piece.

The distribution of projections and photographs across the three rooms gives beholders opportunities, as they move through them, to make their own associations between parts and, in particular, with the three-screen installation in the middle room which includes all the images from the outer galleries. Beholders are also left free to decide how to approach the three-screen projection. By using this format, a range of different kinds of combinations of images becomes possible. Sometimes three shots of the same subject from different angles are placed side by side. Images, however, continually shift from one screen to another, and sometimes two images from one location are combined with a third image from elsewhere. Occasionally the focus is closed down to one image with the other two screens blank, but for much of the installation there is a flow of imagery across the screens that resists being read in a narrative way. The result is highly fragmented and often offers choices about what and how to watch. I am calling this dancerly because of its similarity to the experience of watching dance. This fluid, complex layering of images gives a particular intensity to the experience of watching the installation.

The third aspect I have identified is the vividness of the colour in the films and photographs. The photographs are not printed on paper, but are large-format transparencies displayed in shallow, backlit frames. The overall effect of these, when seen alongside the projections, is one of rich, almost painterly luminosity. Most of the material shown in the projections was shot using 35mm colour film and then transferred to DVD format. Compared with digital media, film allowed Julien more flexibility to capture atmospheric qualities, particularly in the landscape locations. Particularly memorable is an extraordinarily beautiful sequence that shows Myrie and the dancers on white-chalk cliffs beneath a deep-blue Mediterranean sky. The vivid, luminous use of colour in *Western Union* also adds an intensity to the beholder's experience.

There is a documentary-like quality, which Nash mentioned, to the film projected in the first gallery that shows a tracking shot along a fence, behind which the remains of brightly coloured but decayed and rotten small boats are piled. These are traditional fishing boats in which people traffickers dispatched the migrants across the Mediterranean at its shortest point from Tunisia or Libya to Sicily. Designed for inshore sailing, they have often been swamped on the open sea with deadly consequences. The Arabic names painted on their prows point to their North African origin, while their large number suggests the huge numbers of people who have attempted to make this journey. The three-screen installation includes footage in which Julien restaged scenes of migrants in a boat at sea with a group of actual migrants now living in Palermo. This sequence also suggests a documentary mode. There is also a shot of bodies wrapped in silver foil lying on a beach with, in the background, holiday makers. There are stories about drowned migrant bodies being washed up on tourist beaches, although whether this has actually happened is disputed (see McMahon 2005).

In contrast to these documentary-like scenes, are filmed sequences of dancing that open up very different kinds of meanings. In the last gallery, a projection screen shows two male dancers, Alexander Varona and Riccardo Meneghini, who appear and slowly roll and slide up and along a ledge in the Turkish Steps. This is a beautiful chalk cliff at Agrigento in southern Sicily that descends in wide, weathered steps down to the sea. As the dancers roll in a sensitively released way, white traces of chalk rub on to their skin and clothes. In a related sequence shown in the three-screen installation, one of these dancers carries the other across his shoulder along a chalk ledge. The dancer being carried is passive and limp, a close-up of his dangling arm and hand shows them swinging in an easy, relaxed way. All the dancers in the film are dressed in weathered, distressed costumes that suggest the role of migrants. The notion that someone is carrying the body of a drowned migrant is complicated by the fact that Varona is Cuban, while Meneghini, on his shoulders, is white. Later Varona carries a passive Meneghini on his shoulder through the baroque splendour of the Gallery of Mirrors in the Palazzo

Valguarnera-Gangi. These dance images allow Julien to create associations between the migrants, the sea shore and the Palazzo that operate outside any narrative or documentary levels of signification.

Palazzo Valguarnera-Gangi was used in Lucino Visconti's 1963 film *The Leopard* as the setting for the lavish, languorous ball that takes up the film's last 45 minutes. Julien often uses paintings, sculptures or obvious references to well-known feature films in his work in order to make complicated interventions within cultural discourses. For his 1992 film *The Attendant*, for example, Julien filmed Professor Stuart Hall contemplating the 1840 painting *Scene on the Coast of Africa* by the French painter Auguste-François Biard. The film invites the spectator to consider how this eminent Professor's response to the painting might differ from that of the French public in 1840 when slavery was still legal in French colonies. The Gallery of Mirrors functions in the same way in *Western Union*. Like *Ellis Island*, the Gallery is a site of memory where present performed events are coloured by traces of past ones. When Myrie appears, on one screen, in the Palazzo fanning herself and walking through the salon (see Figure 13), while on an adjacent screen Varona carries Menenghini across the same room, the juxtaposition raises questions about cultural values. It asks what meanings the

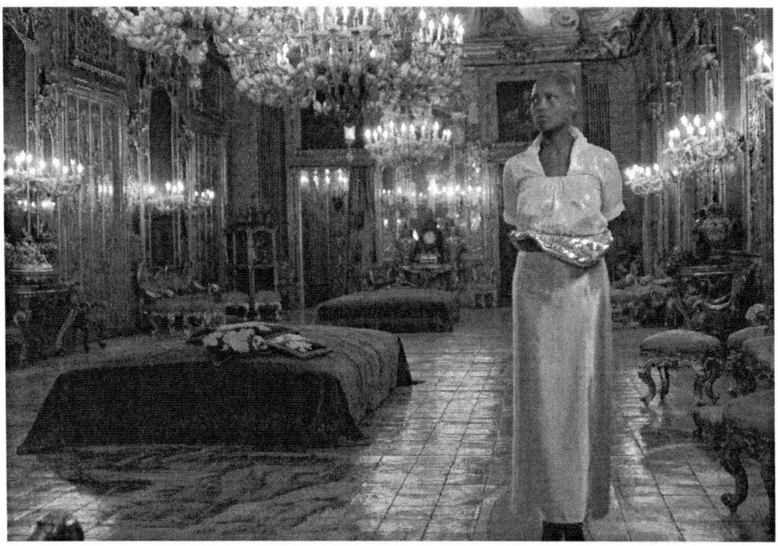

Figure 13 Isaac Julien's *Western Union: Small Boats*. Rebecca Myrie in The Gallery of Mirrors

Palazzo and Visconti's film, and the European high culture that they exemplify, may have for beholders who are black. By interrogating representations in this way, Julien's film uses a specific site to make spectators aware of the kinds of meanings that, in Flusser's terms, migrants bring into play when they interact with settlers.

After Myrie has crossed the Gallery, Varona appears dancing while lying on the baroque painted-tile floor. Spinning round with dynamic force, he bends and folds his trunk, tossing back and forth with urgent intensity. Almost immediately, this image is replaced by underwater film of a dancer twisting, turning and reaching out vigorously around him. It suggests the struggles of a migrant who has fallen overboard. Juxtaposed with the figure in the Palazzo, however, it takes on a broader range of meanings. Both are images of comparable vitality. Regardless of who is drowning or why, a fundamental human response values all signs of life. By doubling this image of struggle through transposing it to the Palazzo, the installation abstracts the dancer's mercurial energy and transforms it into one of the positive qualities that, in Flusser's terms, migrants possess.

Varona's dancing on the Palazzo floor is filmed in slightly out of focus shots from a moving camera that seemingly cannot keep up with him. His movements, and those of the dancer underwater, fill the space around the body with movements that resemble the qualities of contact and release dance. Dancing constitutes another kind of site of memory. Postcolonial theorist Paul Gilroy (1995) has argued that, within African American culture, music and dancing have served as key sites for embodied memories that resisted and survived the dehumanising effects of captive migration and slavery. Varona's solo in *Western Union* resembles solos in other recent films and installations that Julien has made, all of which were danced by dynamic, mercurial black or ethnically marked male dancers. These include Ralph Lemon in *Three* (1999), Ben Ash in *Vagabondia* (2000), Javier de Frutos in *The Long Road to Mazaltán* (2000) and Stephen Galloway in *Fantôme Créole* (2005). Varona's dance movements in *Western Union*, as I have described them, are not in the least African in their technical basis but could, in Brenda Dixon Gottschild's terms, be called Africanist. The dramatic speed and flexibility of Varona's movements exemplify what Gottschild calls high-affect juxtapositions (1996: 14–15).

Dancing also functions as a site of memory in *Ellis Island*. Towards the end of the film there is a long scene where various migrants from different countries all dance together. The music they dance to is a waltz tune, composed by Collin Walcott and arranged and sung by Monk and her vocal ensemble. The dance steps themselves sometimes suggest the middle-class, nineteenth-century waltz and sometimes older folk dance material. The informality of the choreographic structures and the aesthetic sensibility of this 'waltz' recalls the work of Judson Dance Theatre of which Monk was a younger member. Some dancers firmly tread out the waltz rhythm in a pedestrian, late twentieth-century way. Some couples waltz with a self-conscious awkwardness that recalls Monk and Ping Chong's chaotic waltz in their 1972 piece *Paris*. Some dancers in peasant smocks link their arms in a chain, and step sideways in a style that recalls Balkan folk dances. Meanwhile, children skip through the crush of dancers in a headlong race. The camera records this changing scene from a fixed position, and different filmed sequences dissolve one into another. This suggests that different people become involved at different times in an ever-changing dance that was already going on before they arrived and will continue after they have left. Flusser said that the migrant does not cut off all relations with others but weaves connections in co-operation with them, and this is what I suggest happens during the waltz. Rather than blandly assimilating different cultural traits, the 'waltz' imagines a utopian mixing up of all those differences.

I have shown that there are figures who watch and observe in both *Ellis Island* and *Western Union*. These provide the beholder with a point of entry into the works and thus to the underlying themes and issues that they raise. By doing so, the beholder is offered opportunities to form a relation with migrants. As I noted earlier, Flusser proposed that the migrant is both 'a window through which those who have been left behind may see the world and the mirror in which they may see themselves, even if in distortion' (2003: 14). In *Ellis Island*, settlers may see themselves reflected while *Western Union* suggests how the world might appear from a migrant's point of view. In my discussion of each work, I have drawn attention to the way their non-verbal aspects – the singing in *Ellis Island*, the colour in *Western Union*, and the dancing in each – generate affective intensities. I have argued that, in *Western Union*, this shifts the work's significance beyond the particular political circumstances out of which it has arisen towards broader ethical considerations. I suggest that something similar takes place in *Ellis Island*. Both works raise questions about how one relates to others, particularly across the divide separating migrants and settlers. Flusser argued that, because migrants are without mystery, they make settlers aware of their own mysteries. The question he suggests the migrant poses to the settler is:

> How can I overcome the prejudices of the bits and pieces of mysteries that reside within me, and how can I break through the prejudices that

> are anchored in the mysteries of others, so that together with them we may create something beautiful out of something that is ugly?
>
> (Flusser 2003: 15)

Ellis Island and *Western Union* both prove, I suggest, that it is possible to create something beautiful out of something that is ugly, not just at the level of their surface aesthetics, but by reminding both settlers and migrants of their potential for finding ways of relating to each other.

Notes

1. This essay has been adapted and extended from a paper about *Ellis Island* and *Three*, another work by Julien, which I gave at the 2007 CORD Conference *Choreographies of Migration: Patterns of Global Mobility*. This was itself taken from an earlier unpublished essay on the theme of dance and human rights. While writing this initial essay I was very grateful for detailed feedback from Naomi Jackson.
2. The programme was a PERFORMA Commission with Sadler's Wells for PERFORMA07 at the Brooklyn Academy of Music 25th Next Wave Festival. It was co-commissioned by Dance Umbrella (London) and produced by PERFORMA, New York, and Sadler's Wells Theatre, London.
3. The way the film footage was edited for the video installation is different from the edit for the stage version. Email, Gwen Van Spijk, 21 April 2008.

11

Sensation and Memory in Emilyn Claid's *Remember to Forget* (2003) and Gilles Deleuze's Discussion of Francis Bacon's Paintings[1]

Valerie A. Briginshaw

Introduction

Remember to Forget (2003) is the third of a trilogy of works made by Emilyn Claid as part of the writing and performance-making research project entitled *Embodying Ambiguities*.[2] One of Claid's key concerns in making the piece was to investigate the relations between identifiable narratives and abstract movement in dance language by drawing on research carried out in the two earlier works in the trilogy. Of these, *Shiver Rococo* (1999) addressed detailed abstractions of dance language, and *No Bodies Baby* (2002) used narratives and relationships between performers to address a space/time dynamic framework. Claid's investigation of the play between narrative and abstraction resulted in an intense violence of sensation in the movement material performed and her references to other works in the trilogy involved plays with memories. Other memories, such as those of Claid's previous works, of different movement vocabularies in the performers' bodies, and of events that conjured emotions and sensations for Claid and the performers, were also sources, as were the ways in which this piece challenged spectators' memories of presence in performance. There was a recognition that however 'abstract' a stylised dance vocabulary might be, in itself it would harbour shadows of past-performance stories, hence the embodied ambiguity of plays *between* abstraction and narrative, intertwined with and generated by memories.

As writer on the project I had the privilege of witnessing at close hand the process of making *Remember to Forget*. For me, this process closely resembled that of a painter layering paint on canvas and then rubbing out and paring down the surface to reveal glimpses of what is underneath only to build up more layers from which new relations emerge. Gilles Deleuze, writing about Francis Bacon's painting, has suggested that the new relations that result from such a process are 'completely different from relations of resemblance'. Instead they 'produce a more profound resemblance, a nonfigurative resemblance' (Deleuze 2003: 158). For me, the 'nonfigurative resemblance' Deleuze mentions, something that both does and does not resemble a figure, resonates with the results of plays between narrative and abstraction fired by memories in *Remember to Forget*. It is an example of what we termed 'embodied ambiguity'. In his book, *Francis Bacon: The Logic of Sensation* through his philosophical reflections on Bacon's painting, Deleuze develops a 'logic' of sensation; a way of thinking about sensation, that assists me in discussing the violence of sensation in Claid's *Remember to Forget*.

Sensation at one level is the way we experience the world; through our senses, which are crucial for our perception and understanding of what is around us. Senses, sensing, sensation make possible our engagement with the world and with others. Art can foreground sensation, revealing how the senses operate, how the body senses. Art, by materialising and exploring sensation, can also reveal its potentiality; the possibility, for example, of new sensations, of new ways of engaging with the world and with others. This gives it political potential. Deleuze's discussion of sensation in Bacon's painting, his ways of thinking about sensation, open it up in this way. By foregrounding its potential to act on us, to affect us, as well as our potential to sense or experience sensation, new possibilities for sensation are suggested.

Discussing painting and sensation, Deleuze claims 'there are two ways of going beyond figuration', or beyond narrative, and these are: 'either toward abstract form or toward the Figure'. He cites Cezanne who, he claims, named 'this way of the Figure: sensation' (2003: 34). Deleuze expands, indicating that: 'The Figure is the sensible form related to a sensation; it acts immediately upon the nervous system, which is of the flesh' (ibid.). Deleuze, again invoking Cezanne, claims sensation is located in the body in painting, not just in the body as an object, but 'as it is experienced as sustaining *this* sensation' (ibid.: 35, his emphasis). There are obvious resonances with dance here. The claim that the body not only experiences, but sustains sensation suggests that sensation lives in the body. Furthermore, its direct transmission to the nervous system, the flesh, foregrounds its intensely, visceral or violent impact. This is the violence of sensation that, I am suggesting, became evident in *Remember to Forget*, because Claid too was moving beyond figuration or narrative toward the Figure.

Deleuze claims that when sensation, 'acquires a body through the organism, [it] takes on an excessive and spasmodic appearance, exceeding the bounds of organic activity. It is immediately conveyed in the flesh through the nervous wave or vital emotion' (2003: 45). This excessive, spasmodic appearance, which exceeds the bounds of organic activity, results in what Deleuze terms a body without organs. This is what he means by the Figure. This body without organs is liberated from the organs or shackles which fix it. It is no longer an organism which 'imprisons life' (ibid.: 45), trapped in or grounded by its organic structure. This freeing up of the body means it can 'give an account' of how 'every sensation implies a different level' and how sensations move 'from one level to another' (ibid.: 48), so that, for example, 'what is a mouth at one level becomes an anus at another level' (ibid.). Bodies without organs, that are released or emancipated in this way, are evident in *Remember to Forget*. The violence of sensation we experience as a result, in the process, helps us think bodies and sensation; what it means to sense and engage with the world differently.

Whilst witnessing the process of making *Remember to Forget*, parallels with Deleuze's discussion of Bacon's painting became evident in the notions of: 'presence/absence', 'layering' and 'deformation'. These three different approaches to aesthetic production cut through normative conventions to unleash the violence of sensation common to both artists' work. They are explored in some detail here because of the ways in which they can open up the workings of sensation, revealing the role that memories often play.

The body without organs which becomes sensation in Deleuze's discussion of Bacon's work also becomes the Figure. The Figure, for Bacon and Deleuze, in its non-narrative, non-abstract sense, has resonances with the plays between narrative and abstraction that are at the heart of Claid's *Remember to Forget*. These plays, for Claid, are bound up with the performer's seductive engaging with **absence** as **presence** that is created from a process drawing heavily on memories. For Deleuze, 'presence' is everywhere in Bacon's paintings and, similar to sensation, acts 'directly on the nervous system' (2003: 51). It is my contention that the way Claid uses memories in the body to create a seductive absence/presence in performance resonates with Bacon's treatment of the figure to produce the kinds of non-figurative resemblances that Deleuze discusses.

Layering, whether of paint on the canvas over scrubbed out areas in Bacon's paintings, or of movement material invested with memories in Claid's choreography, is another parallel concern. Both artists are concerned with embodying multiples in their work, and their processes reveal this. As Deleuze claims for Bacon's paintings: 'Every sensation and every Figure, is already an "accumulated" or "coagulated" sensation' (2003: 37). The 'plurality of constituting domains' (ibid.), with which Deleuze characterises sensation, is also evident in the layering of memories evident in *Remember to Forget*.

For Deleuze, 'force is the condition of sensation' (2003: 56) and invisible forces, when exerted on bodies in Bacon's paintings, result in what Deleuze terms **deformation**. Deleuze claims 'sensation is the master of deformations, the agent of bodily deformations' (ibid.: 36). Bodily deformations in Bacon's paintings, according to Deleuze, are the result of sensations passing or moving from one level or domain to another. Similar bodily deformations, I am claiming, are also evident in the disjunctions and grotesque movement material in *Remember to Forget*. They liberate the Figure, which is sensation.

Remember to Forget

Remember to Forget resulted in a 45-minute performance, with music commissioned from Stuart Jones, the composer of *Shiver Rococo,* and lighting from Charlie Balfour. Like *Shiver Rococo* and *No Bodies Baby* before it, it was multidimensional, performed in an oblong space with spectators seated on all sides, with gaps between them so that the dancers could perform behind as well as in front of them. Like its predecessors, it had a cast of five: three female and two male, one of whom, Martin Welton, also performed in *No Bodies Baby.*[3] The show opens in darkness with Martin entering at one end and spotlit. Looking around him, he begins to speak: 'At first it went forwards I remember remembering that [...] this morning running into this afternoon. I still have clear memories of the slow segue of yesterday into tomorrow.' Martin's texts – there are five in the course of the performance – are like the movement material in the way they veer between narrative and abstraction, and play with memories and sensations. Martin is inhabiting the same persona he created for *No Bodies Baby.* At one level the piece can be seen as Martin's memories, brought to life by the other performers who inhabit his imagination, and realised in part through the sensations he experiences.

As he begins to talk, a shaft of white light extends from him to the other end of the space, where two pairs of performers are entering. They are accompanied by sparse, eerie, occasional echoey, night sounds and groans half-human, half-animal heard in the distance. Patricia and Alicia are perching on the backs of Grant and Sarah who are slowly crawling forward. All are dressed simply in black, close-fitting trousers and tops. As the pairs move diagonally outwards into the half-light, gradually advancing towards Martin, energies seem to be released from them. They are Figures in the sense that each 'body exerts itself in a very precise manner, or waits to escape from itself in a very precise manner' (Deleuze 2003: 15). They appear full of potential, poised for action. Sitting high and proud on the backs of their now standing partners, Patricia and Alicia slowly turn and look around them with their arms spread in front, as if surveying vast lands from on high. They seem like distant dignitaries from another time, another place. Their slow turnings affectively trace contours conjuring memories that hover between them as Figures full of intensities.

As the lights fade, to sounds of water trickling, all five performers disperse, slowly disappearing into gaps between the audience where, spot lit, bending forward with hands in front of their faces, they appear to be guiding imaginary emissions of liquid or words from their mouths to the ground, letting them ripple through their fingers. These figures are intensely affective, not only because of their precise, unusual actions, but specifically because of their positioning in and between spectators in the half-light. As audience, we can sense some performers very near to us, maybe next to us or along the row from us, whilst others we glimpse opposite. All are releasing this imaginary outpouring from their gaping mouths. These are Figures, in Deleuze's terms, bodies without organs, bodies attempting to escape from themselves through one of their organs. They have resonances with the scream in Bacon's *Head VI* (1949), which Deleuze describes as: 'the operation through which the entire body escapes through the mouth' (2003: 16). The unlikely actions of these Figures in *Remember to Forget,* in their excessiveness and their proximity, unsettle us. They suggest new ways of experiencing bodies. Their silent 'scream-breaths' are the sensations produced as the result of dismantling 'the organism in favor of the body, the face in favor of the head' (ibid.: 45). As the water sounds fade we hear high-pitched single notes, strained, vibrating strings and occasional echoey whistles. The air of nocturnal mystery remains but with fuller sounds now than at the beginning. The lighting, now filling the space, casts strong shadows as one by one the performers enter.

Alicia is Spanish, short with a trim figure and jet black hair pulled back from her face. Wandering into the space looking slightly disorientated, she lunges forward. Her body seems to fill with confidence. Then balancing on one leg, she slowly turns, curving her arms around an imaginary lost form in front of her. She is embracing an interval full of intensities, of memories. Grant, a tall, imposing New Zealander, his arms framing his head in balletic style, is mirrored by Patricia, a boyish-looking, half-Nigerian, half-German young woman. She turns as Grant walks past her, her body sensing the energy of his. She breaks her classical pose by reaching further with one arm and arching her head back to look up. Alicia runs to Grant and we see Sarah waltzing around the periphery without a partner. Patricia runs to Martin and lifting his arms at the elbows, places them on her shoulders before moving back and lowering them. Alicia runs to the corner with arms out to grasp Sarah, who has gone. With no-one to hold, she turns instead to Martin.

One brief encounter after another occurs. Nothing is established in terms of relationships or movement material. Instead, we can only glimpse fragments of what might have been. Hints of intimacy – Patricia placing Martin's arms on her shoulders; indications of loss – Sarah's partnerless waltzing or Alicia's frustrated reaching towards her; and memories – Grant's and Patricia's balletic poses. These mysterious figments are harbingers of what is to come. In these opening moments we glimpse material, which we see extended later, from various sources of memories. We also see plays between abstraction and narrative – Alicia's seemingly lost character who transforms into a ballet dancer – and between different movement languages – ballet, ballroom dancing, contemporary dance. The importance of sensation in touches, whispers and looks might also be triggered by, and trigger, memories. The work's title, *Remember to Forget*, and Martin's opening speech, articulating the workings of memory, further suggest the conundrums and ambiguities associated with memory.

As the piece progresses, changes in the music and lighting states coincide, announcing different 'scenes', and the five performers become a little more established. Martin is set apart from the others. He is the only one who speaks, and he moves awkwardly compared to the rest. He occasionally breaks into extremely athletic routines, hurling his body into the air, running fast around the periphery and jumping repeatedly as he did in *No Bodies Baby*. When discussing Bacon's painting, Deleuze indicates that it is the task of art to 'render visible forces that are not themselves visible' (2003: 56). He mentions elemental forces like weight and gravity. Martin's affective athleticism, his incessant running round the periphery close to the audience and his repeated jumping make the forces of weight and gravity visible. We witness close up his body continually and forcefully being grounded by gravity. Martin is, at times, the centre of attention, and he begins and ends the piece. Sarah is often in attendance: following him, carrying him, cradling him, perching on his back. She seems like a material memory he is unable to shift. Yet she also has encounters with the others. She appears to occupy an ambiguous space between him and them. She is white, of medium height with short, straight dark hair. She is a shadowy figure, sometimes barely visible, often lurking in the background around the edge of the space, watching the others.

Alicia, Patricia and Grant are often seen dancing together in pairs and trios. Their classical training is evident, their limbs lusciously cutting swathes of sweeping gestures, and grandiosely extending in long elegant lines. Balletic bodily memories seem to nostalgically haunt them. They become Figures in the sense that their bodies appear overtaken by these memories. Bursts of energy seem to appear as quickly as they go, surprising us with their fleetingness. Their encounters are sometimes lively, energetic and competitive. Counter-tension is often evident when adjoined limbs are extending – arms linking or a hand grasping an ankle. But there are also softer moments of delicate intimacy when Patricia elusively keeps slipping out of Grant's grasp, seemingly both close and distant simultaneously. Or when Alicia and Grant, wrapped around each other, are slowly exploring their bodies, improvising with eyes closed.[4] Bodily memories guide their sensing.

An air of mystery is created throughout by the strong lighting states, which change from shadowy and dim, to stark white or misty blue. The music, too, is powerfully evocative: swelling strings, baby cries and bird song join weird groans and straining cello sounds. It is nostalgic, suggesting dreams or memories of elsewhere. There is no single, linear narrative. Rather, we are confronted with performers encountering each other, bodies sensing, energies being released, creating affects, interweaving with texts, all bound together in a complex web of sensations interlaced with memories.

Seductive 'presence/absence' – between narrative and abstraction

Memories triggered sensations when making *Remember to Forget*. These sensations vividly haunt the performance, often confronting the audience violently, given their physical proximity to the dancers. For Deleuze and for Claid, sensation is closely bound up with notions of 'presence/absence' and seduction. Deleuze claims presence is 'the first word that comes to mind in front of [...] Bacon's paintings'. He continues: 'Interminable presence. The insistence of the smile beyond the face and beneath the face. The insistence of a scream that survives the mouth, the insistence of a body that survives the organism [...] And in this excessive presence the identity of an already-there and an always-delayed' (2003: 50–1). This notion of presence characterised by an oscillation between 'an already-there and an always-delayed' has resonances with Claid's notion of the 'oscillating play of the illusive surface and real body of a performer's presence' (2006: 91). For her, this characterises 'the seductive relations in the conventions of ballet' (ibid.). In more radically innovative dance like *Remember to Forget*, however, Claid claims: 'the dancer seduces her spectator by a reverse strategy. The real/illusion strategy for seduction is turned inside out' (ibid.: 94) Instead of watching the surface illusion for glimpses of the real body as in most ballet, in *Remember to Forget* the spectator searches: 'the real body for glimpses of the performed surface of illusion – which is present only in its absence' in the spectator's memory. The spectator is seduced by 'an absence of presence' (ibid.: 95). They engage with absence as presence. The resonances with Deleuze's discussions of presence in Bacon's paintings are evident in the seductive processes. Plays between what is absent and present, there and not there, known and not known, are seductive. For Deleuze, the value of painting is that it 'directly attempts to release the presences beneath representation, beyond representation' (2003: 52). I am claiming these 'presences' are different to the notion of performance presence that Claid discusses, which is more akin to Deleuze's notion of 'representation'. In his book *Difference and Repetition*, Deleuze (1994) discusses the limitations of representation: its grounding tendencies and relations with origin, the ways in which it fixes and constrains thought. He welcomes the potential art has for the 'abandonment of representation' (ibid.: 68–9). For Deleuze, presence in painting, and for Claid, the seductive absence as presence in performance, have powers to release and play between what Deleuze calls the smile beyond and the smile beneath, and the already-there and the always-delayed, and what Claid terms the real body of the performer and her/his illusive surface image. These seductive plays between them disrupt the limiting tendencies of representation, of the sign languages that often constitute narratives. These plays allow for an oscillation between narrative and abstraction involving sensation and memory.

When working with performers on *Remember to Forget*, Claid used the terms 'full body/empty body' to suggest this complex play between different bodily states. She describes this process in some detail in her book, *Yes? No! Maybe...*(2006). Referring to the making of the penultimate episode, 'Portrait', of her solo for herself, *Virginia Minx at Play* (1993), Claid writes that the creative challenge 'was to find the place on my body where fullness of memory and emptiness sparred with each other on a fine line of tension.' (2006: 155). She describes working with a series of autobiographical memories where she: 'relived the gestures and rhythm of the breathing repeatedly, finding every detail, every turn of head and angle of chin, storing them as kinaesthetic memory, as gestures that felt connected to the experience.' She then recreated 'the body language of the experience with every muscle, eye movement, angle of head, and placement of fingers and toes as physical task.' Finally, she ends, 'I emptied my body of emotion, leaving the empty shell [...] I worked to find that fine line between content and form, balancing in the space between full body and empty' (ibid).

In other words the body was both full and empty at the same time – physically full of all the details required to reconstruct the memory, but emotionally drained such that the experience was not being re-enacted. This is not a simple distinction between physical fullness and emotional emptiness. It is much more complex. It requires the performer to produce affects whilst keeping the full details of how they are produced from the spectator, leaving an open, empty space or surface for the play of the spectator's imagination and desire. An example of the effect of this 'full body/empty body' practice in *Remember to Forget* – the title itself is instructive – is when Patricia is walking whilst tortuously reaching upwards with every muscle and sinew of her body. This phrase was crafted from a childhood memory of Patricia's. Claid had set the performers the task that, above, she describes having set herself. Patricia's walking is *full* of graphic physical description, the violence of sensation, yet it is not a narrative or resemblance of her memory. It is, in Deleuze's terms, 'a more profound resemblance, a nonfigurative resemblance' (2003: 158). The violence of the different levels of sensation at play in her body reveals an accumulation or coagulation indicating the complexity and inevitable plurality involved in such a Figure.

Deleuze describes figuration as the process in painting which 'recovers and recreates but does not resemble, the figuration from which it came'. He continues: 'the act of painting is always shifting, it is constantly oscillating between a beforehand and an afterward' (2003: 98). This is evident in Patricia's walking. Another example was derived from the 'Portrait' section from *Virginia Minx at Play*, which Claid used to illustrate the full body/empty body process, where she is making a journey from young girl to old woman. Claid drew on this source whilst working with the performers playing between images and sensations of a small baby and an old woman. This resulted in the performers folding and unfolding, rolling and reaching. Dancing that was often awkward, slightly off-balance, encapsulated the ambiguities of an 'aging baby'. Accompanied by crying baby sounds, groans and sighs, the effect was strange and disconcerting in its accumulation of levels of sensation and memories of both a small baby and an old woman. It was always shifting, constantly oscillating between a beforehand and an afterward, suggesting new ways of sensing.

Discussing the play between narrative and abstraction, Deleuze refers to Proust who, he claims:

> did not want an abstract literature that was too voluntary philosophy, any more than he wanted a figurative, illustrative, or narrative literature that merely told a story. What he was striving for [...] was a kind of Figure, torn away from figuration and stripped of every figurative function: a Figure-in-itself.
>
> (2003: 67)

Many of the Figures in Bacon's paintings and in *Remember to Forget* are figures-in-themselves in this sense. They are neither figurative – illustrative or narrative – nor figural or abstract, but in between. They also resemble Proust's description of the operation of involuntary memory discussed by Deleuze who claimed: 'it coupled together two sensations that existed at different levels of the body, and that seized each other like two wrestlers, the present sensation and the past sensation, in order to make something appear that was irreducible to either of them [...] this Figure' (ibid.: 67). This description closely resembles the full body/empty body practice, which also concerns wrestling between a present sensation and a past sensation, inevitably engaging the workings of memory.

Layering

Another process that contributes to the plays between abstract and narrative is layering. Deleuze often refers to Bacon's various scrubbing, brushing, rubbing and wiping off techniques, where the thickness of paint is spread in different ways to 'make invisible forces visible' as asignifying traits (2003: 5, 21, 58). These contribute to processes of decomposition and recomposition. (ibid.). Claid also assembled thick layers of movement material whilst making *Remember to Forget* only to arrange and rearrange them and pare them down such that hints of forces, energies and sensations remained in performers' bodies but were not consciously articulated. For example, towards the end of the process of making the piece, when Claid had taken out many of the original 'thick' layers of material, she would often say to performers 'show me what you used to do here'. She would then ask them to retain a small hint of what had been there before, either actually in movement terms, or more often in a memory of a bodily sensation. Through this decomposition and recomposition process, she was making some of the invisible forces of memory visible. The result is that, as Deleuze claims for Bacon's painting: 'several forms may actually be included in one and the same Figure, indissolubly, caught up in a kind of serpentine, like so many necessary accidents mounting on top of one another' (2003: 160). When this layering process occurs in *Remember to Forget,* it reflects the workings of memory in a performance hovering between narrative and abstraction. It opens up possibilities of seeing more than one thing at a time in one performance, of seeing embodied ambiguities, where bodies are continually sensing and becoming. These pluralities suggest new, more complex ways of sensing that have political implications. As Deleuze claims, appearances are superimposed 'one on top of the other, in ways different to those vouchsafed to us from life' (ibid.: 180). An example of this occurred in the process of making *Remember to Forget* when Claid would remove a performer from a duet or trio, but ask those remaining to perform as if that performer were still there. It is, as Deleuze asserts with Bacon's paintings, as if 'spirits [...] haunt the wiped off parts' (2003: 21). Thus one series of trios between Grant, Patricia and Alicia was reduced to a series of duets where each performer in turn, when absent from the original trio, witnessed what remained of it in the other two dancers' duet. The effect was like a palimpsest. Seeing the spaces that they inhabited, where now only memories remained, also had an effect on the dancers' watching and witnessing. They were there in the trio and not there, visible and invisible simultaneously.

The dance vocabularies within Claid's and the performers' memories were also layered within the piece. As Deleuze indicates for painting: 'it would be a mistake to think that the painter works on a white and virgin surface. The entire surface is already invested [...] with all kinds of clichés, which the painter will have to break with' (2003: 11). The clichés of dance vocabularies in *Remember to Forget* were often broken literally in performance, as unexpected 'accidents' occurred with performers releasing energy through deformations and disjunctions.

Deformations

When creating *Remember to Forget,* Claid worked with the dancers establishing lines in the body, often derived from classical ballet vocabularies. She then introduced disjunctions or deformations where a different kind of energy was released breaking the lines, disarticulating the bodies. For example, when Alicia and Patricia are supporting each other in a strong counter-tension pull with opposing arms joined, Claid encourages them to prolong the tension to an extreme until they are forced to break the line. This changes the composition of their bodies and the torsion in the interval between them. It is as if, as Deleuze describes, 'an intense movement flows through the whole body, a deformed and deforming movement that at every moment transfers the real image onto the body in order to constitute the Figure' (2003: 19). The affect of this, for the spectators, is unexpected and disorientating. It suggests thinking of bodies and sensations differently.

Often in *Remember to Forget*, as in the two earlier works in the trilogy, the lines in dancers' bodies are not only broken, but, as Deleuze suggests in painting, they are 'split, diverted, turned in on [themselves], coiled up and extended beyond [their] natural limits', resulting in grotesque and deformed imagery, in 'disordered convulsion' (2003: 46). For example, at one point Martin and Sarah are leading Patricia and Grant in a grotesque coupling that became known as a 'monkey walk'. Sarah and Grant are walking forward on their haunches, reaching up to clasp the hand of their partner leading them. In the shadowy half-light, their distortion in time and space, playing between narrative and abstraction, evokes memories. These are evident in the tension between the energies released and the grossly moving, animal-like forms. They appear like Figures exceeding their normal organic forms, as bodies without organs. They resemble some of the animated, distorted figures from Bacon's paintings where bodies become entangled with one another, metamorphosing into grotesque couplings. Deformation and distortion accumulates in the dance when Sarah and Grant, whilst rising up and placing their heads in their partners' stomachs, force them backwards in another contorted perambulation.[5] These bizarre wracked, wrestling forms, through rendering visible the invisible, suggest new possibilities for sensation. Through distortion and deformation, Deleuze suggests, a painter makes visible 'the original unity of the senses', making a 'multisensible Figure appear' (ibid.: 42). These dancing Figures, I am arguing, are also multisensible.

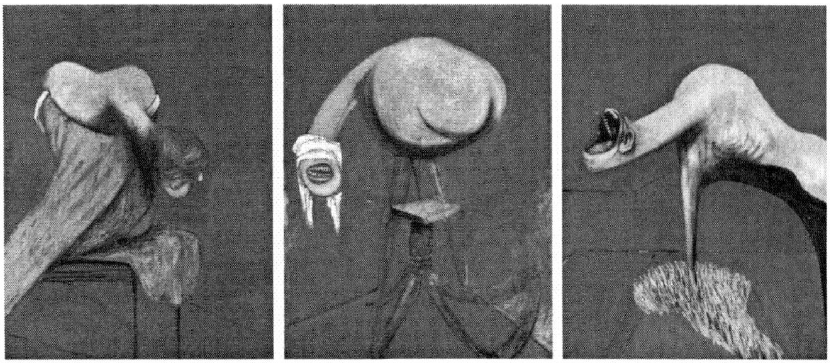

Figure 14 Francis Bacon's *Three Studies for Figures at the Base of a Crucifixion* (1944)

Courtesy of Tate Images.

The grotesque and deformed couplings in both *Remember to Forget* and Bacon's paintings often appear monstrous. Another particularly vivid example from Bacon's work that reminds me of some of the contorted postures and gestures that constitute embodied ambiguities in the trilogy of Claid's works is his *Three Studies for Figures at the Base of a Crucifixion* (1944) (see Figure 14). Deleuze refers to this painting as a 'triptych of monsters, where the head with bandaged eyes (central panel) is not a head preparing to die, but an abominable head that smiles along the horizontal deformation of the mouth' (2003: 75). When this deformed smile in the central panel becomes a ghastly gaping hole threatening to bite in the right-hand panel, it resembles the 'aging baby' dancing from *Remember to Forget*. It echoes the interval when the performers, slowly turning their heads, open their mouths wide and sink their teeth into the flesh of one of their forearms. This graphic, cannibalistic gesture, a memory from an earlier work in the trilogy, *Shiver Rococo,* (see Chapter 3, Figure 4) acts directly on the nervous system. It reveals the complexities of the intensities of sensation in a multisensible Figure.

Conclusion

Without wishing to suggest that Claid's work actually drew on Bacon's paintings, there are nevertheless resonances between *Remember to Forget* and the qualities Deleuze identifies through his discussion of these paintings. Each can open up facets of the other. Bacon and Claid both explore the force of sensation which underlies these resonances and the plays in that elusive space between narration and abstraction inhabited by memories. Throughout *Remember to Forget*, the force of sensation is profoundly prevalent because the audience surrounds, and is sometimes surrounded by, dancing which is full of intensity produced by a powerful combination of sound, lighting and the energies of performing. The action of invisible forces on the body, which Deleuze claims constitutes sensation in Bacon's paintings, is also palpably present in the performers' seductive absence of presence, layered memories and the forces of disjunction and deformation. This could not result without the detailed crafting of layers of material or without the continual investigation and play, in Claid's case between full body/empty body, involving memory, wrestling between a past sensation and a present sensation.

I suggested at the outset that through exploring sensation and memory in *Remember to Forget al*ongside Deleuze's discussion of Bacon's paintings, potentialities for new ways of sensing and of engaging with the world and with others might be revealed. I am arguing that in *Remember to Forget*, the ways in which sensation and memories are vividly and pluralistically experienced and felt in Figures point to new ways of engaging with our senses. As Deleuze claims for Bacon's paintings, the dancers' performances 'in the very act of "representing" horror, mutilation, prosthesis, fall or failure, have erected indomitable Figures, indomitable through both their insistence and their presence. They have given life a new and extremely direct power of laughter' (2003: 62). Laughter can be tremendously liberating, because, like the kinds of sensation discussed above, it is often a result of direct action on the nervous system. What has become evident from this exploration of *Remember to Forget*, aided by Deleuze's reflections on Bacon's paintings, is that sensation is complex and multi-levelled. It affects the body in a direct and immediate manner without mediation or translation. If this is the case with painted Figures, I am suggesting, the intense experience of sensation when the Figures are suggested by actual, live, dancing bodies might be even more immediate. For, as Deleuze claims, when 'the visible body confronts the powers of the invisible, it gives them no other visibility than its own. It is within this visibility that the body actively struggles affirming the possibility of triumphing' (ibid.: 62). This possibility of triumphing is positive, suggesting potential. I am claiming that this potential is political since new ways of engaging with the world and others through our senses are implied. This positive potential is evident in art such as Bacon's paintings and Claid's *Remember to Forget* because of the affective impact of sensation imbued with memories.

Notes

1. An earlier version of this chapter was given as part of a keynote address to the Nordic Forum for Dance Research at their Annual Conference in Reykjavik, Iceland, in 2004 and this was published in the conference proceedings.
2. See www.embamb.com and also Chapter 3 in this volume which focuses on *Shiver Rococo,* the first in the trilogy of works that resulted from the *Embodying Ambiguities* project.
3. The cast of *Remember to Forget* were Grant McLay, Patricia Okenwa, Alicia Herrero Simon, Sarah Shorten and Martin Welton.
4. A memory of Lisa and Yalckun's 'blind duet' in *Shiver Rococo* – see Chapter 3.
5. A memory of heads being placed on stomachs in *Shiver Rococo* – see Chapter 3.

Section 6
Reimagining Dancing Together

Introduction

The two chapters in this section are the most recent ones we have written, both discussing dance works from the last decade. They are 'open works' because their radical exploration opens up the movement potential of the dancing body, extending the kinds of affective relations that it can generate. These works challenge fixed ways of thinking about theatre dance and offer beholders opportunities to experience performance differently. Their openness enables what we suggest can be seen as a performative ethics. This arises from the new ways their performance brings people together and because of the practices of the self that we suggest they propose. Valerie's discussion of Jonathan Burrows and Matteo Fargion's *Both Sitting Duet* points to the ways in which differences and repetitions in the choreography are productive, freeing it from the confines of normative representation. She argues that, as a result, the beholder can witness affective relations between the two men which suggest new ways of relating to an Other in an ethically hospitable manner. Through a discussion of performative presence, Ramsay asserts that the kinds of practices of self proposed in recent works by Yvonne Rainer and Xavier Le Roy exemplify similar relations between affective openness and ethics. He argues that Rainer's work, in effect, celebrates the individual's potential for liberation from hegemonic norms, while Le Roy critiques power relations within the institutionalised dance world. By so doing, both choreographers propose ethical practices that can become sites of resistance against dominant ideologies. In both chapters, our discussions of these performances point to the potential for alternative, more ethical relations with others where we can reimagine dancing together in utopian futures to come.

12

Affective Differences and Repetitions in *Both Sitting Duet*[1]

Valerie A. Briginshaw

Introduction

When I saw Jonathan Burrows and Matteo Fargion's *Both Sitting Duet* (2003), I was immediately intrigued by the ways in which these two men were playing with rhythmic patterns, specifically by repeating and differentiating them, in such a way as to suggest new possibilities for relationships; between ideas or thoughts, and between two white, apparently heterosexual, males. Their plays with repetition and difference in the performance, for me, resonated with some aspects of Gilles Deleuze's theories in his *Difference and Repetition* (1994). Here I explore *Both Sitting Duet* and *Difference and Repetition* alongside each other. I am not *comparing* the two works, although at times it may appear so, but rather juxtaposing my responses to each of them and playing between these. In the process I explore certain facets of each and unravel elements of the mysteries they entail.

Through exploring Burrows and Fargion's dance work alongside Deleuze's ideas, it becomes possible to see the repetition in the dance providing a framework for the new relationships that the piece suggests. Deleuze's theories in *Difference and Repetition* help to suggest why this is so and what kinds of new relationships are possible. The potential for change that becomes apparent is evident in possibilities for two sorts of new relationships. The first is between thoughts of difference and repetition, which can be applied to various contexts within and beyond the field of dance studies. Ways of *rethinking* what we mean by difference and repetition become apparent. The second is in the new possibilities for relationships between two men that are suggested.

Both Sitting Duet and *Difference and Repetition* are very different works from different domains, with different practices and cultures – dance and philosophy – nonetheless, certain parallels between them are evident, which underpin the resonances I perceive. They are both, importantly, open works: they open up respectively movement, dance and music, and thought, ideas, concepts and philosophy. Through their focus on, and different explorations of, difference and repetition, both works investigate and play with relations between parts. It is not insignificant that the starting point for *Both Sitting Duet* was a piece of music by Morton Feldman entitled *For John Cage* for, as becomes clear, Feldman's philosophy underlying his composition, which focuses on relations between parts, also concerned *rethinking* difference and repetition.

I am analysing the dance at the level of an 'extended repetition'. Deleuze distinguishes the difference in kind between extended repetitions, which have volume, take up space and are actual, and 'intensive repetitions', which are immaterial, do not have volume and are virtual. As becomes clear, there is no given concept or notion of *Both Sitting Duet* governing this analysis. I see it rather as an open and fluid entity and my aim is to foreground the transgressive potentials of this extended repetition for making differences that matter between two men or between self and other. I also want to clarify that Deleuze's encounter with difference differing, as becomes clear, is a risky encounter for any embodiment and hence for dance. Within dance it is important to keep sight of some necessary limits to Deleuze's recurrent differings in order to retain a notion of an embodied dancing subject. This is an instance where Deleuze's theories, if taken to the extremes he at times suggests, do *not* resonate with my reading of the dance.

Both Sitting Duet is a 45-minute piece devised and performed by longtime collaborators: dancer and choreographer, Jonathan Burrows, and composer, Matteo Fargion. As the title indicates, the duet is largely sedentary. The two performers sit on chairs close to and slightly turned towards each other, near to and facing the audience. Large open notebooks on the floor in front of each contain their scores and are referred to throughout, the performers occasionally turn a page, although rarely at the same time. The duet consists of rhythmic, repetitive patterns of mainly hand movements often touching other body parts such as thighs, chest and the other hand, and occasionally also, the floor. The ways in which these patterns are developed, varied, contrasted, performed in unison, overlapped and alternated, constitute the multiple differences and repetitions we witness, as do the range of rhythms, dynamics and qualities played with; from regular to irregular, fluid to fierce, vigorous to gentle, and 'throw away' to carefully placed. The concept of counterpoint is relentlessly explored, played with and in the process exploded.

Deleuze claims that 'repetition belongs to humour and irony; it is by nature transgression or exception, always revealing a singularity opposed to the particulars subsumed under laws' (1994: 5). The importantly transgressive character of Deleuze's notion of repetition, which does not repeat the same, but reveals singularities that, in their opposition to 'the particulars subsumed under laws', can shift our thinking, is paramount in his argument. This has similarities with the affective, and at times comic, elements of *Both Sitting Duet*, which can also be seen as transgressive and subversive because of the ways in which they trouble expectations of what is normally thought acceptable behaviour between two straight men. The relaxed intimacy and familiarity is a departure from the norm. The complex humour of *Both Sitting Duet* often results from the idiosyncrasies of the relationship between the two men, which is evident in looks, timing and the material repeated and differentiated between them. Their relationship resonates with Deleuze's argument in *Difference and Repetition* where it concerns relations with 'the Other'. For Deleuze, the Other is bound up with notions of individuation and difference. The role of the Other allows individuations to take place. The particular individuations in the extended repetitions of *Both Sitting Duet* suggest seeing encounters with others otherwise. The Other, for Deleuze, is also bound up with simulacra, which, as copies of copies, are intimately involved with repetition. *Both Sitting Duet*, in its plays between the two performers, between music and dance, and between repetitions and differences within movement and sound, creates and plays with copies of copies. It can be seen as a series of simulacra.

After introducing *Both Sitting Duet*, I focus on the resonances between it and Deleuze's theories. These are played with in no particular order since the journey is nomadic and open-ended. It meanders and spirals in a rhizomatic way such that there is repetition but, as in the 'nomadic distributions' of Burrows and Fargion and of Deleuze, it will always involve difference and never be the same.

Both Sitting Duet

In the Preface to *Difference and Repetition*, Deleuze asserts: 'I make, remake and unmake my concepts along a moving horizon, from an always decentered center, from an always displaced periphery which repeats and differentiates them' (1994: xxi). Continuous movement where nothing is ever fixed is important for Deleuze's philosophy. For him, thoughts, ideas, concepts are forever on the move. It might therefore be claimed that any dance, because of the movement involved, would have resonances with his work. However it is not just 'a moving horizon' that concerns him, but one that is from 'an always decentered center, an always displaced periphery' (ibid.). Decentring and displacement shift a viewpoint away from the norm, and defamiliarise things such that we are able to see them differently. I am suggesting that, because of the distinctive performance of Burrows and Fargion in *Both Sitting Duet*, we see relations between two men differently. Such disabling mechanisms as decentring can suggest the unexpected and things left to chance. Although Burrows and Fargion follow a score, so nothing appears to be left to chance, *Both Sitting Duet* seems impossible to fathom. The work looks deceptively predictable, but it surprises throughout. As with Deleuze's concepts, the patterns in the duet seem to be made, remade and unmade but 'from an always decentered center and an always displaced periphery' (ibid.), we never know quite where they are coming from. This element of uncertainty is perhaps one of its attractions. Within dance the 'centre' can be seen in a style of movement or phrasing that is predictable and recognisable from a certain aesthetic, such as classical ballet or traditional modern dance within Western theatre dance, for example. A displaced 'periphery' is evident when such predictable familiar forms are toppled or subverted such that the *status quo* is threatened, which is what I am claiming for *Both Sitting Duet*. There are significant parallels with Feldman's characterisation of his music composition. He claims that he contributed to 'a concept of music in which various elements (rhythm, pitch, dynamics, etc.) were *decontrolled* [...] this music was not "fixed"' (my emphasis) (2000: 35). He also writes of 'constant displacement' of, for example, a 'rhythmic shape' within composition and of deterring 'the natural propulsion' of the music (2000: 142). These are further examples of decentring and displacement strategies, which throw the expected progression of a phrase or element out of kilter and, because of this, importantly enable new ways of seeing and experiencing things. They provide a 'view from elsewhere'.

Both Sitting Duet consists mainly of abstract hand gestures performed to no music by two ordinary looking men. This is another parallel with Deleuze's work: the abstraction in the piece results in a lack of obvious meaning, identity or point of reference. There seems to be no point to what Burrows and Fargion are doing since they are not representing anything with their gestures. In *Difference and Repetition* Deleuze discusses at length the limitations of representation, its grounding tendencies and links with notions of origin. He claims that difference 'cannot be thought in itself, so long as it is subject to the requirements of representation' (1994: 262), and when discussing repetition, he suggests 'a complete reversal of the world of representation' (ibid.: 301). *Both Sitting Duet also* works against notions of representation and origin with its weird gestures that do not appear to mean anything. Even if one does recognise a move such as a 'thumbs up' sign, it is repeated and played with amongst other signs and gestures such that it loses its point of reference. These gestures seem to have no origin, they represent nothing beyond themselves. Similarly Feldman writes of giving up controls in his musical composition such that the musical elements 'lose their initial, inherent identity' (2000: 35).

The duet is deceptively complex. It looks comparatively simple but is exceedingly intricate and difficult to fathom. On several levels it embodies simultaneously, in a Deleuzean manner, various 'couplets' made up of very different components. Its performers are, on the one hand, ordinary or unspectacular – they are dressed plainly and often perform straightforward, pedestrian movements, such as slapping their thighs. On the other hand, they are extraordinary or spectacular – their deft performance of intricate patterns of hand gestures, perfectly timed, is virtuosic. They are both dependent – watching each other carefully, picking up on and responding to the other's performance – and independent – getting so absorbed in the nuances of their own performance that at times they seem oblivious of the other's presence. Often these apparently dichotomous characteristics coexist in the performance. It is *both* pedestrian and virtuosic, spectacular and unspectacular and ordinary and extraordinary, and all at the same time. In *Both Sitting Duet* the boundaries between music and dance are repeatedly blurred. Alongside beautifully constructed designs of hands dancing, we hear unusual sounds and rhythms, as hands smooth across clothed bodies, or fingers flick, scrape or knock on different surfaces, resulting in, as one spectator commented, a confusion of the senses (see Brown 2003a: n.p.).

The piece opens as Burrows and Fargion walk into the performance space to sit down. They are simply dressed in jeans and boots. Burrows wears a long-sleeved T-shirt, whilst Fargion wears a shirt, both with rolled-up sleeves. They sit, hitching up their jeans for comfort, adopting a typical male pose with feet astride and knees apart. Throughout, they often place their hands on their thighs in a typically masculine posture. Sometimes they lean forward in this pose as if they are about to start a conversation, like two men in a bar or pub, but instead they surprise us by turning a page of their score or taking the bend forward into another phrase of movement. This exemplifies the plays between pedestrian and virtuosic moves, which characterise the work. It is also an example of decentring or displacement. The recognisable, familiar pose that resembles two men in a bar or pub is defamiliarised or decentred by what follows it. Sometimes it is difficult to know whether the pair are performing or not. There are many pauses; when one, or other or both appear to be resting or marking time, just sitting with hands on thighs or in laps. One watches the other, or stares into space or at the audience, only to break into a flourish of elaborate activity. The boundaries between performance and non-performance are often blurred in this way and because the pauses vary in length, the resumption of activity, or its cessation, often catches us unawares. This is another example of displacement that has come from Feldman's original composition, on which *Both Sitting Duet* is based. In his essay 'Crippled Symmetry', Feldman writes of 'disproportionate' (2000: 135) or asymmetrical symmetry, he claims that he likes working with 'patterns we feel are symmetrical' and then presenting them in particular contexts, for example, using the device of 'a longish silent time frame that is asymmetrical' (ibid.: 140). The length of the silences or pauses changes and never becomes predictable.

The repertoire of movement material in *Both Sitting Duet* seems limited, because it is mainly focused in the hands and arms, but also extensive, because of the variations employed. The piece begins with Fargion flicking the backs of his fingers down his thighs, then tapping his thighs quite sharply with the outside of his hands, and raising his hands over his ears without touching them. Almost simultaneously Burrows gestures with both hands diagonally down whilst picking up an invisible fleck of dust from the floor with his middle finger. Each repeats their moves six or seven times almost simultaneously, and then Burrows picks up Fargion's phrase and performs it with him a few times. Next they alternate this and finally Burrows resumes his earlier pattern whilst Fargion continues his. By now Fargion's phrase has been repeated around 20 times and Burrows's slightly less. In parts the relationship between their hand patterns is like a conversation; with one making a statement and the other responding, but at other times both 'speak' at once, sometimes 'saying' the same things, sometimes 'saying' different things. At other times they appear to follow or imitate one another so that it becomes less like a conversation and more like a game where the rules keep changing. Each repetition is a little different: sometimes one will look at the other or his hands; sometimes one finishes slightly before or after the other; or the energy invested or the size varies. The differences can be hardly noticeable, or blindingly obvious when the material or manner of performance changes. This is typical of what follows. Both then work at various patterns; sometimes the same, sometimes different, sometimes simultaneously, sometimes successively. Disparate repetitions almost become competitive in a friendly manner and exemplify the pair's close relationship. Although the two men do not speak, rarely touch each other or catch each other's eye, they seem attuned to each other.

As the piece progresses, more adventurous phrases are introduced. At one point the pair appear to be throwing something over their shoulders, at another, they vigorously swing one arm forward as if tenpin bowling. The concentration and energy that each puts into these seemingly meaningless tasks, is sometimes humorous, possibly because pedantic attention to detail is often combined with a casual 'throw away' approach. Ironic stark contrasts occur between everyday moves, like counting the fingers one by one, and more theatrical emotional flourishes when hands are shaken and crossed in front of the face as if to say, 'No! no! no!' When Burrows has rapidly repeated a thumbs up sign, a circle made with finger and thumb, and a flat palm 'Stop' sign seemingly endlessly, he looks into the space above him as if trying to remember something. Is he trying to recall this relatively simple phrase, or what comes next, or something completely different, perhaps? The effect appears parodically 'laid back' after some of the dynamic activity that has preceded it.

The performers seriously approach each task in a workmanlike fashion. They look like labourers or craftsmen: carpenters planing wood; potters shaping clay; bakers kneading and folding dough; but none of these actions is a direct mimed copy that can be identified. They are all played with through repetition and variation. So, although at times the semaphore resembles that of cricket umpires, traffic policemen, bookies or orchestra conductors, it is in fact none of these. It has been played with and repeated so many times out of context that any original source is no longer evident. Gestures may contain hints of recognisable codes, but they become abstracted through: incessant repetition; combination with something different; or changes in design, size or dynamic. Productive rather than reductive repetition, involving playful differentiations within and between dance and movement, music and sound and one performer and another, are the hallmark of *Both Sitting Duet*.

Repetition

In *Difference and Repetition*, Deleuze asserts that 'the subject dealt with here is manifestly in the air' (1994: xix). It has currency and contemporary relevance. I also believe *Both Sitting Duet* is of its time. Its subversive and transgressive, affective tendencies, evident in new ways of being, or becoming, in Deleuze's terms, have the potential to suggest ways of rethinking relationships between notions of repetition and difference and between two men. One review mentioned that the 'imaginative dancemaking' reminded of the 'potentialities of life itself' (thedanceinsider.com cited in Hutera 2003: n.p.). Deleuze, referring to 'modern life' writes of the perpetuation of 'mechanical and stereotypical repetitions, within and without us' (1994: xix). There is a sense in which we are trapped in an entropic space of repetition of the same because we fail to recognise the potential for movement, fluidity and change in repetition that is imbued with difference.

In another review of *Both Sitting Duet*, Ann Williams comments, 'there's nothing that could truly be described as "dancing" in "Both Sitting"' (2003: n.p.). In Deleuze's terms, she is reiterating 'mechanical and stereotypical repetitions' of the same concept of 'dancing'. By implication she restates traditional boundaries that contain and fix ideas about what 'dancing' is and can be. She has failed to recognise the potential for change in the concept, which *Both Sitting Duet* manifests. The performance shows how, in Deleuze's terms, 'dancing' can be an open Idea, which embodies difference and excess, rather than a concept, which is confined and closed. In Deleuze's words, 'Ideas are not concepts; they are a form of eternally positive differential multiplicity, distinguished from the identity of concepts' (1994: 288). Williams provides further evidence of her singular focus on 'concepts', in Deleuze's terms, by commenting on the different performances of Burrows, 'a dancer', who has 'unmistakable grace' to which 'the eye is continually drawn', and of Fargion, 'a composer, not a dancer, yet he matched Burrows movement for movement with only slightly less ease and elasticity' (2003: n.p.). Here boundaries between dance and music, which have been displaced in the performance, opening these concepts up as Ideas, are re-erected in writing through a repetition of the same. There *are* some subtle differences between the executions and physicalities of Burrows and Fargion as I discuss below, but my point here is to illustrate Deleuze's notions of 'concept' and 'Idea' in action. Williams's review is underpinned and bound by 'concepts' of dance and music that are 'fixed' in the Deleuzean sense, whereas *Both Sitting Duet* exemplifies Deleuze's notion of the 'Idea' which he sees as open, fluid and differentiated and characterised by repetition which is imbued with difference.

Deleuze has developed his notion of the Idea in opposition to concepts which he sees as limiting because of the ways in which they tie thought to notions of essence, identity, representation or signification, which derive from rationalism and the dominance of the rational unified subject in the history of Western philosophy. As he asserts 'rationalism wanted to tie the fate of Ideas to abstract and dead essences [...] to the question of essences – in other words to the "what is X?"' (Deleuze 1994: 188), whereas 'the events and singularities of the Idea do not allow any positing of an essence' (ibid.: 191). There are parallels here with Feldman's claims for his music not being 'fixed' and his suggestions that 'as controls are given up [...] these elements [of rhythm, pitch, dynamics, etc.] lose their initial inherent identity' (2000: 35). In Deleuze's terms, like his Ideas, they are 'extra-propositional and sub-representative' (1994: 267) or developed in 'sub-representative determinations' (ibid.: 155).

The resonances between Deleuze's Idea and *Both Sitting Duet* are particularly apparent in the ways in which both exemplify combinations of opposites. I have suggested that the performance of *Both Sitting Duet* is characterised by couplings such as pedestrian/virtuosic, spectacular/unspectacular and dependent/independent. A coupling which Deleuze uses for Ideas is 'distinct-obscure' as opposed, he argues, 'to the clear-and-distinct of Apollonian representation, Ideas are Dionysian, existing in [...] a pre-individuality which is nevertheless singular: the obscure zone of an intoxication which will never be calmed; the distinct-obscure as the double colour with which philosophy paints the world' (ibid.: 280). This tendency for ideas to encapsulate doubling and to be consequently unbalanced, or in Deleuze's terms, ungrounded, keeps them dynamic and differentiated or individuated, continually on the move, and full of multiplicities. I see these characteristics in *Both Sitting Duet*. I would say for *Both Sitting Duet*, as Deleuze claims for Ideas, that it liberates difference and causes 'it to evolve in positive systems in which different is related to different, making divergence, disparity and decentering so many objects of affirmation which rupture the framework of conceptual representation.' (ibid.: 288). We see this in the ways in which the various hand patterns continually differ, never repeating the same, but metamorphosing like the patterns in an ever-changing kaleidoscope and also, importantly, in the ways in which the two men relate to each other. The openness, divergence and decentring of this difference differing can 'rupture the framework of conceptual representation' because it defamiliarises and shifts expectations in terms of what repetition and relations between two men can mean. This is what renders Deleuze's philosophy and *Both Sitting Duet* transgressive and subversive with the potential to bring about changes in thought or ways of rethinking repetition and difference and relations between two men.

When discussing repetition, Deleuze asserts: 'the task of life is to make all these repetitions coexist in a space in which difference is distributed' (1994: xix). It is my contention that *Both Sitting Duet* goes some way toward opening up such a space, not least because its format of a series of repetitive patterns performed mainly with hands enables a focus on the distribution of subtle but multiplicitous differences. For example, Burrows and Fargion complete a simple phrase turning their palms out to face the audience, whilst the backs of their hands are placed on their knees. It resembles the response of a small child when asked what it is holding. These straightforward repeated actions are performed almost in unison at a moderate pace such that we are able to see the subtle differences between them, which come from the two performers. Fargion is more stolid, broader and more firmly set, his hands have a weight about them that Burrows's lack. They look slightly bigger, whereas Burrows is able to achieve more lightness and delicacy in his gestures. These differences are almost imperceptible but because the actions are repeated, we begin to glimpse them. The phrase is accelerated and Burrows begins his gesture forward from the shoulders so that it is larger, it becomes less pedestrian and more theatrical with more style and volume. Here the development of the phrase has further opened the space in which difference is distributed in Deleuze's terms. It has resonances with Deleuze's notion of Ideas which as 'multiplicities' of 'differential elements' (ibid.: 278) 'shine like differential flashes which leap and metamorphose' (ibid.: 146). It is also one of the many examples of the ways in which *Both Sitting Duet* differentiates difference. However, contrary to Deleuze, in *Both Sitting Duet* the difference is never differentiated to an extreme or to infinity, there is always a limit brought about by a change to the next pattern, or by the end of the series, or the end of an instance of the performance.

The repetition here is imbued with difference, but normally repetition is thought of as being concerned with reiterating something of the same. This is why in this instance, according to Deleuze, we need to rethink what we mean by difference and repetition. This requires 'two lines of research' (1994: xix). One is to argue for and fashion a concept of 'difference without negation' (ibid.: xx), that is to rethink difference such that it does not have to involve 'opposition' and 'contradiction'. *Both Sitting Duet* blurs boundaries between 'oppositional' or contradictory notions, such as the ordinary and the extraordinary, performance and non-performance and the dependent and the independent, providing examples of the coexistence of differences that do not negate each other.

The other line of research, for Deleuze, is to conceive of repetition in which 'bare repetitions (repetitions of the Same) would find their *raison d'être* in the more profound structures of a hidden repetition in which a "differential" is disguised and displaced' (1994: xx). Many different levels of repetition are at work in *Both Sitting Duet,* some more hidden than others, where 'differentials' are disguised and displaced. In addition to the choreographic repetitions within the piece, there are repetitions compared with previous performances and rehearsals, from other works by Burrows, such as *Hands* (1995),[2] and from already existing repertoires of movement. Plays with disguise and displacement of difference within repetition at these various levels can result in unexpected elements, which shock and surprise. For example, there is a phrase of vigorous arm swings that start going backward, which Burrows and Fargion begin performing in unison. Just as we are getting into the infectious rhythm of this pattern, Fargion suddenly stops and rests his hands on his knees – he misses a swing such that when he resumes he is swinging back when Burrows is going forward. Then Burrows stops suddenly, and when he restarts the two are in harmony again. But then each of them keeps stopping, and they go in and out of time but not in any apparent pattern. Here the 'differential' between Burrows and Fargion, in Deleuze's terms, keeps being displaced, such that we do not know where to expect it next. There is an element of play, deception or disguise at work, which makes the repetition productive. There is a sense in which, as Deleuze asserts: 'repetition is this emission of singularities, always with an echo or resonance which makes each the double of the other, or each constellation the redistribution of another' (ibid.: 201). There are parallels with his notion of the Idea which is concerned with division 'a capricious, incoherent procedure which jumps from one singularity to another, by contrast with the supposed identity of a concept' (ibid.: 59).

Deleuze claims that repetition is everywhere: 'it is in the Idea to begin with, and it runs through the varieties of relations and the distributions of singular points. It also determines the productions of space and time [...] In every case repetition is the power of difference and differentiation' (ibid.: 220). The ways in which repetition is played with in *Both Sitting Duet* suggest some of its powers of difference and differentiation. When Burrows and Fargion simultaneously wipe their hands over their faces, they appear to be removing perspiration. Where Burrows wipes his hand down once, Fargion wipes his face three times with alternate hands, his rhythm is brisker. Initially his three hand wipes appears to measure Burrows's one hand wipe exactly, but then Burrows makes his last longer, so Fargion has to wait for him to finish before he repeats his phrase. Each time, although they perform the face wipes together, the timing is slightly different, such that either Burrows finishes with Fargion slightly after him, or a little later. Repetition's powers of difference and differentiation are played with and demonstrated. 'Varieties of relations', between the face wipes, in this instance, can also be seen, as can 'distributions of singular points'; here, the points where Burrows finishes his phrase, which vary each time. As with many of the plays between Burrows and Fargion's performances in *Both Sitting Duet*, the divergences or differences are evident both spatially and temporally. Spatially, we see Burrows's hand return to his lap after his face wipe, either alongside Fargion's or slightly behind his, or on its own. Temporally, the hands complete their phrase at the same time, or successively; one is delayed and behind the other's time, sometimes slightly and sometimes considerably. As Deleuze indicates, repetition can be seen to 'determine the productions of space and time' (ibid.).

An important element of Deleuze's argument concerns the distinction between repetition and resemblance. He asserts 'repetition and resemblance are different in kind – extremely so' (1994: 1). This distinction provides the space for repetition to diversify, to depart from the same and engage with difference. The abstract and apparently meaningless character of the gestures performed in *Both Sitting Duet* departs from a logic of identity, resemblance or representation. There are resonances with Deleuze's notion of the Idea which he claims 'is extra-propositional and sub-representative' (ibid.: 267). The gestures for the most part are non-mimetic, they do not resemble or represent anything but themselves. They do not obviously refer back to some recognisable identity. They behave like simulacra.

Simulacra

Simulacra are the result of rethinking repetition without notions of origin or identity. The authenticity of an original is undermined, paralleling Deleuze's notion of the Idea which he also develops to oppose thoughts of origin, identity, concept, and by extension the cogito or 'thinking subject'. For Deleuze 'simulacra are the letter of repetition itself' (1994: 17). They play an important role in his discussion of difference and repetition because, as he indicates, 'all identities are only simulated, produced as an optical "effect" by the more profound game of difference and repetition' (ibid.: xix). Simulacra are more than just copies or imitations, because their repetition involves difference. They challenge notions of copying or imitating, which involve resemblance. As Deleuze claims: 'the simulacrum is not just a copy, but that which overturns all copies by *also* overturning the models' (his emphasis) (ibid.: xx). Simulacra are copies taken to extremes. They are copies that, through repetition of copies of copies, involve difference such that the original model is no longer evident. In Deleuze's terms, it is overturned. *Both Sitting Duet* consists almost entirely of series of copies of copies that involve differences of differences that simulacra constitute.[3] Burrows and Fargion happily reiterate one repetition after another. They appear to be copies or imitations, but they are not, because they revel in the plays of differences of differences they produce.

Deleuze suggests modern art as a site of simulacra claiming, 'art is simulation, it reverses copies into simulacra' (1994: 293). He cites Andy Warhol's 'serial' series as an example where, 'Pop Art pushed the copy, copy of the copy, etc., to that extreme point at which it reverses and becomes a simulacrum' (ibid.: 294). The series of repetitions and differences, which are rife in *Both Sitting Duet,* are similar to those in Warhol's works. They are instances of the law of diminishing returns where, in Walter Benjamin's (1973) terms, each repetition diminishes the value or 'aura' of the original, so that through repetition it becomes no longer an original and a copy, but a series of copies without an original. In *Both Sitting Duet* the repetitions of the vocabularies of gestures and movements lose the affective power they might have if only seen for the first time, but become affective in a different sense through the patterns and textures of their repetitions, just as with Warhol's prints of Marilyn Monroe's face, for example.

References to different dance forms in *Both Sitting Duet* are examples of simulacra because the actions repeated out of context lose their original point. Burrows performs the classical *port de bras* of five arm positions whilst Fargion accompanies each with a slow hand clap. When repeated, the claps' timing alters relative to Burrows's arms. The two men's performances, when combined, as copies of copies, become something else. They are simulacra, disconnected from the (flawed) Idea of representing models. They parallel the sub-representational character of Deleuze's Ideas. Amongst a welter of small hand signs in the duet, occasional Indian classical dance *mudras* are glimpsed, but it is impossible to identify them as they merge into the simulacral mirage of signs being repeatedly performed. When Burrows and Fargion join hands, with arms up and out in front, a minuet is momentarily suggested, but, because the pair are seated, looking down at their scores or elsewhere as they repeat the gesture, it is removed from its original source. It has become a simulacrum through the plays of repetition and difference that have occurred.

Both Sitting Duet is evidence of simulacra at work in several senses. Feldman's score, *For John Cage*, from which the piece originates, which Burrows and Fargion decided not to name in the programme, is a homage, like Warhol's prints of Monroe, and a simulacrum in itself. It is a violin and piano piece, which Burrows and Fargion admit they had been 'obsessed with' (in interview with Hutera 2003: n.p.). When asked why they did not name it, Burrows responded: 'because those people who know it would always be waiting for it, and those who don't would feel excluded. It's kind of irrelevant' (ibid.). Burrows's sentiments here resonate closely with Deleuze's theories concerning the limitations of notions of origin and identity, which are bound up with those of resemblance and representation, which constrain thought. This is another instance of *Both Sitting Duet,* because of these sub-representational tendencies, paralleling Deleuze's notion of an Idea. Deleuze cites the simulacrum as 'the act by which the very idea of a model or privileged position is challenged and overturned' (1994: 69). Again he refers to art claiming that 'when a modern work of art develops its permutating series and its circular structures, it indicates to philosophy a path leading to the abandonment of representation' (ibid.: 68–9). This is precisely what *Both Sitting Duet* does. Through its 'permutating series and its circular structures' (ibid.) it suggests the abandonment of representation, in the sense of reference to origins, because, as Burrows indicates, on that level it's irrelevant.

Feldman's score itself and what Burrows and Fargion did with it are examples of how simulacra operate. According to Burrows, the 'constant small difference within repeating patterns is very much part of the philosophy of what Feldman was doing' (2004). Apparently he was inspired in his late work by 'the small shifts in pattern and the colour of the dye in [...] beautifully hand-made Oriental rugs' (ibid.), which can be seen as further examples of extensive repetitions in Deleuze's terms, also operating like simulacra. Burrows continues, 'the technique Feldman used to create these changes within the repetitions [...] was to write the rhythms in a more complicated way than necessary. This means that even in a simple sounding loop of notes, the musicians are always translating and counting, which means they're never sure, and they never fall into the step of marching' (ibid.). The complexity of differing differences in Feldman's score is what makes it a series of simulacra, which are no longer concerned with representation, resemblance and a repetition of the Same. As Deleuze indicates elsewhere, 'the simulacrum is constructed around a disparity, a difference, it interiorises a dissimilitude' (1990: 258).

Burrows and Fargion made a direct transcription of Feldman's score 'with the same tempo bar for bar, note for note' (Burrows in interview with Hutera 2003: n.p.). But as Burrows indicates, although he and Fargion 'knew the value of [the complexity within] Feldman's repetitions', when they began working on *Both Sitting Duet*, they 'overlooked the importance of this rhythmic device to break them up' (2004). Burrows claims 'we just simplified all of Feldman's complex counting. But then [...] we discovered for ourselves the reason he'd written it that way, and we had to find our own technique to break the rigidity of the repetitions and breathe life into them again!'(ibid.). In the process, I suggest, Burrows and Fargion, in Deleuze's terms, became aware of how much differing the differences mattered to break the rigidity of repetitions of the Same.

They did this by creating simulacra: by creating series of differing repetitive and rhythmic movements and sounds, and, in the process, by also each creating their own scores. These scores, left on stage available for view after the performance, consisted of combinations of numbers, words, hieroglyphics of various kinds, such as dashes and squiggles, and musical notes and time signatures. Each of these was another copy of a copy, a simulacrum. When Burrows and Fargion looked at what they had created, using only the repetition from the Feldman score, they claimed they were 'quite surprised, because whereas the world of this music is a kind of hovering, rocking, quiet thing, we seemed to be more jolly and folk-dancey' (in interview with Hutera 2003: n.p.). Here it seems, in Deleuze's words, are: 'rebellious images which lack resemblance', or which, like the Idea, are 'sub-representational' – 'simulacra' (1994: 272). The resulting relationships of the Feldman score to *Both Sitting Duet*, and of Burrows's score and performance to Fargion's, are, in Deleuze's terms, like 'divergent stories unfold[ing] simultaneously, it is impossible to privilege one over the other [...] the one story does not reproduce the other, one does not serve as a model for the other: rather, resemblance and identity are only functional effects of that difference which alone is originary within the system' (1994: 125). Burrows and Fargion have transported us from the 'hovering, rocking, quiet' world of the Feldman score, through repetition and difference via copies of copies, simulacra, or different 'stories', to another different potential world, a 'jolly', 'folk-dancey' one.

As is evident in *Both Sitting Duet* simulacra are transgressive and subversive. They are, from one of Deleuze's perspectives, condemned because of their 'oceanic differences [...] nomadic distributions and crowned anarchies' (Deleuze 1994: 265), and because they are 'ungrounded false claimants' (ibid.: 274). If taken to extremes, Deleuze's notions of 'oceanic differences [...] nomadic distributions and crowned anarchies' leave no room for an embodied dancing subject. Within dance, we need limits on these nomadic processes of difference endlessly differing. Nevertheless glimmers or glimpses of 'nomadic distributions' and 'oceanic differences' can be perceived in the multiple differences that occur in *Both Sitting Duet*. In Deleuze's terms, they 'emerge' from 'groundlessness' (ibid.: 276) and in their rebelliousness and flightiness they are variously: 'phantasms' (ibid.: 126, 127), 'demonic images' (ibid.: 127), 'dreams, shadows, reflections, paintings' (ibid.: 68). Deleuze suggests the 'world of simulacra' is one of metamorphoses of 'differences of differences, of *breaths* [...] of [...] mysteries' (ibid.: 243, his emphasis). One critic writing of *Both Sitting Duet* alludes to 'Two men on stage', 'an instant story. Brothers, rivals, workmates, lovers, Laurel and Hardy [...] all these evocations emerge like wispy genie' (Brown 2003b: n.p.). Importantly these are, for her, 'evocations' and not representations and their emergence like 'wispy genie' suggests an air of mystery. Similarly, Deleuze writes of Ideas as 'multiplicities with differential glimmers, like will-'o-the-wisps' (1994: 194). The coexistence of the differences evident in these evocations qualify them as instances of simulacra in Deleuze's terms; they appear, disappear and coexist like images in dreams, phantasms, shadows, reflections or paintings.

Because of the systems of differences of differences, of the affirmation of divergence and decentring that Deleuze attributes to simulacra, it is perhaps not surprising that he suggests that if they refer to any model, it is to 'a model of the Other, an other model, the model of difference in itself' (1994: 128). One of the distinctive features of *Both Sitting Duet* is the intriguing relationship that transpires between its two performers. Many reviewers mention that they are long-term collaborators; Fargion has composed for Burrows for 13 years, and many comment on the ways in which an obvious friendship and intimacy between the pair is evident in their performance, which one terms a 'buddy ballet' (Brown 2003b: n.p.). This is evident in looks between the two; smiles; their split-second timing that seems dependent on intimate knowledge of the material and each other's performance; they seem to sense when the next move should begin without looking at each other. All this suggests a relaxed laid-back atmosphere and a sense of ease between the two that suggests intimacy. The various relations we witness between Burrows and Fargion during the course of the performance suggest new possibilities for relationships between two men, in part because they have resonances with some of Deleuze's theories concerning the Other.

The Other

Deleuze uses the term 'Other' in *Difference and Repetition* in various senses, but a key source is Lacanian psychoanalysis. The binary oppositions of subject and object and presence and lack are expressed in traditional Freudian psychoanalytic theory via the Oedipal myth by reference to the 'Other'. Deleuze rejects these theories, which he claims: 'oscillate mistakenly [...] from a pole at which the other is reduced to the status of object to a pole at which it assumes the status of subject' (1994: 260). As a result, he claims, 'the structure of the other, as well as its role in psychic systems, remained misunderstood' (ibid.). The 'Other' for Deleuze is defined by 'its *expressive* value', 'its implicit and *enveloping* value' (ibid., my emphases). For him, 'the Other cannot be separated from the expressivity which constitutes it' (ibid.). It is evident throughout *Both Sitting Duet* in one sense as the differentiating factor that plays within and between Burrows and Fargion and the sequences of movements they execute. When they perform a sequence of five parts reaching up with both arms, circling to the side with one, reaching up again, circling heads with arms, and throwing palms forward, it is impossible to separate the differentiating elements from the expressivity of the sequence as a whole. This is in part because the elements are intertwined with each other when the phrase is repeated, since the order of the parts is changed. In doing this Burrows and Fargion are playing with the repetitive elements. They are doing this by reordering the elements but also by repeating them differently because their manner of repetition is clearly individual or within their own style, they are not repeating them in the normal sense. Trying to make the material look identical or exactly the same does not seem important to them, it does not appear to be part of their agenda. Like the oriental rugs that inspired Feldman, it is the subtle differences in repetition that make the repetition distinctive and that give it the expressivity and enveloping value of the Other in Deleuze's terms. Consequently, whatever the sequence expresses to us is embodied within it in its totality of repetitions and variations, or within the differentiating factor that plays between them. Sequences such as this can be regarded as instances of 'Other-structures' in Deleuze's terms. It is in this way that the repetition imbued with difference of *Both Sitting Duet* provides a framework for new ways of thinking repetition, which in turn provide frameworks for new relationships.

Deleuze sees the Other in terms of 'individuating factors' or those that make a difference. This is similar to his notion of Ideas, which are also 'expressed in individuating factors' (1994: 259). He claims that in psychic systems: 'there must be centres of envelopment which testify to the presence of individuating factors. These centres are [...] constituted neither by the I nor by the Self, but by a completely different structure belonging to the I-Self system. This structure should be designated by the name "other"' (ibid.). The Other 'functions as a centre of enwinding, envelopment or implication' (ibid.: 261). In *Both Sitting Duet* the gestures, patterns and looks exchanged between Burrows and Fargion seem also to be 'centres of enwinding, envelopment or implication'. What is implicated or expressed by them cannot be separated from them, or situated in either one of the performers, it rather exists *between* them. They sit very close to each other, they often perform similar or what appear to be the same gestures, each often looks at the other, and occasionally they look at each other. For example, when the pair perform a simple phrase consisting of placing their palms down on their knees, one palm up the other down and turning them over, they look at each other's performance as if to try and fathom the puzzle or riddle that they are performing. These looks are affective. They add a recognition of the other involving difference to the repetitive plays being performed. These looks form part of a circulation of signs between the two performers and their performance that can be seen as centres of enwinding, envelopment or implication. They make up the piece's affective qualities. As a result, our eyes are drawn to the space between them imbued with these intensities and multiplicities. It is as if a seductive energy emanates from that space, because of the interconnectivity of the Other structure that plays between them.

It is perhaps not surprising that the subject/object, self/other binary is dissolved in a Deleuzean manner in this performance, since Burrows claimed that when they were making the piece their intention was: 'to find something that we could place between us that was neither too much Matteo nor too much me, but which could be an arbiter of our process' (in Hutera 2003, n.p.). They have certainly succeeded in doing this. The 'something' they have found; the 'arbiter' of their process is, in Deleuze's terms, the 'Other-structure'. This makes the work radical. As one reviewer suggested: 'Burrows's *Both Sitting Duet* isn't a usual choreography at all, since it is an equal partnership between him and his friend' (Brown 2003a: n.p.). What makes it distinctive from most other collaborative works is that the process is revealed to us in the performance. Rather than appearing as a finished product, in *Both Sitting Duet* it looks like Burrows and Fargion are improvising or playing as we watch. We see the equality of the partnership in action. It is impossible to distinguish one as subject and the other as object within it. The subject/object divide is dynamic within the dance, rather than static. I am not arguing with Deleuze for total desubjectivisation, for individuation is needed for the purposes of distinction, I am arguing rather for the need to see beyond the subject/object divide. As Deleuze claims, the Other should not 'be anyone, neither you nor I', 'it is a structure [...] implemented only by variable terms in different perceptual worlds – me for you in yours, you for me in mine' (1994: 281). This is what happens in the dance. In the course of *Both Sitting Duet* the relationship between Burrows and Fargion is such that they inhabit each other's perceptual worlds and this extends to an ethics of hospitality and welcome evident between them. These encounters with the Other structure in *Both Sitting Duet* can be seen to be renegotiating it, the dance is 'talking back' to Deleuze's philosophy; the alterity involved demolishes the potential despotism of the structure. Consequently, the dance can be seen as a force for political change, because the potentially problematic iconic images of white, middle-aged, straight males, traditionally associated with the dominant subject position, are repeated differently and transformed through the minutiae of differences that matter in the performance.

In their radical blurring of subject and object and self and other, Burrows and Fargion are expressing a different potential world, where two ordinary looking, middle-aged men can intimately engage in a complex and, at times, delicate undertaking requiring considerable skill and concentration. They are both dependent and independent at the same time. It is this interconnected unusual relation with the other, who is part of the same self/I structure, that makes the piece subversive, transgressive and radical. The performance replaces the iconic imagery traditionally associated with white male subjectivity suggesting ways of seeing that subjectivity differently. In part this is because, as Ramsay comments in his review of the piece: 'the performers' informal, unseductive, uncharismatic presence directs attention away from the dancer towards the movement itself and the affective qualities that their movements generate' (2003: n.p.).

The structure of the Other represents for Deleuze 'the tendency towards the interiorisation of difference' (1994: 261). Instead of seeing things in terms of the binary opposition of self or 'I' with other or 'you', the individuating and differentiating properties of otherness, which for

Deleuze are also seen in Ideas and repetition imbued with difference, fracture the 'I' and dissolve the self as separate entities. Thus the 'I' and self interiorise the individuating factors of difference and become other too. In *Both Sitting Duet*, it is plays with, and performance of, repetition imbued with difference that release the individuating and differentiating tendencies of Ideas to, as Deleuze puts it, 'swarm around the edges of the fracture' (ibid.: 259). The performance results in a relationship between self and other, Burrows and Fargion, two men, which is no longer between self and other understood as separate entities, or subject and object, but rather as similarly differentiated open entities, where the I has been fractured and the self dissolved, each enveloping and expressing something of the other.

There are parallels with the relationship between music and dance in the duet. Burrows asserts, 'we all know what that relationship is, but we can't really grasp it. We think that we dance to music. But [...] that's not what I do, and I don't think that's what I see other people do. I see them hanging and falling always around the music, but never grasping hold of it' (in Hutera 2003: n.p.). 'Hanging and falling always around the music' and 'never grasping hold of it' is excessive and incarnates in Deleuze's terms 'an Idea'. It is an example of a different kind of relation with 'the other': a productive repetition involving difference that is open, expressing new potential worlds. If Burrows and Fargion grasped hold of the music, then they would be repeating the Same, in Deleuze's terms, constrained within a way of thinking bound to representation. Whereas, as Burrows indicates, he and Fargion avoided repetition of the same by trying to find a way to perform 'where we're not marching in step, not like an army going "crunch, crunch, crunch"' (ibid.). In order to avoid repetition of the same – the 'crunch, crunch, crunch' – he and Fargion worked on the piece such that 'the counterpoint between us is somehow in all the spaces around the marching' (ibid.). By being in the spaces around the marching, the counterpoint between Burrows and Fargion imbues the 'marching' with difference. It is another example of an 'Other-structure' which avoids repetition of the same. The counterpoint involves a dependent independence, and as Fargion claims, assumes 'a love between the parts' (ibid.), suggesting a different and productive relation with the Other, which is transformative.

Conclusion

Deleuze suggests that we need to rethink what we mean by 'repetition' and 'difference'. The ways in which *Both Sitting Duet* plays with difference and repetition can change the way we think, because in these plays we witness how repetition can be productive when imbued with difference. Burrows and Fargion's performance also has the potential to change the way we think about relations with an Other. Seeing some of the implications of these complex philosophical theories in practice in the dance performance, opens up Deleuze's ideas. Links between repetition imbued with difference, Ideas that are individuated and differentiated and his notion of the Other-structure become evident in the ways in which, through performing and playing with the repetitive patterns in Feldman's score and making them their own, Burrows and Fargion share and inhabit each other's perceptual worlds.

Seeing the Other in *Both Sitting Duet* as the individuating or differentiating part of a self/I structure, in Deleuze's terms, suggests that our relations with another, or others, can be differentiating forces for change. Through the affective, nuanced repetitions of differences that matter in the dance, we are presented, not with a single model, but with an open series of differences. There is a sense in which these unusual, surprising and, at times, unexpected plays with difference liberate concepts, thoughts and expectations about dance and music and about relations with the Other, from the history of representations that bind them. This gives *Both Sitting Duet* radical potential. The ways in which each of the two performers encounters and engages with difference show how each can inhabit and share the other's perceptual world. In the process, alternative ways of becoming, that embrace difference and supplant the ways of being of the traditional, dominant, white, male subject, are suggested.

Notes

1. Thanks go to Jonathan Burrows, Ruth Chandler, Sarah Rubidge and Ramsay for reading drafts of this chapter and making constructive critical comments which have informed it. An earlier version was published in *TOPOI*, Vol. 24, No.1, 2005: 15–28 under the title 'Difference and Repetition in *Both Sitting Duet*' and material from that is reproduced here with kind permission of Springer Science and Business Media.
2. A solo made for television with music by Fargion, also consisting entirely of hand gestures.
3. See also our discussion of multiples and failed copies in Lea Anderson's *Yippeee!!!* in Chapter 6 in this volume.

13
What the Dancing Body Can Do: Spinoza and the Ethics of Experimental Theatre Dance[1]

Ramsay Burt

During a panel discussion in London in February 2007 chaired by Jonathan Burrows with the choreographers Jérôme Bel and Xavier Le Roy and the dramaturge Bojana Cvejić, the topic of 'conceptual dance' came up.[2] This is a label which some continental European critics have been using to describe recent European dance work, but is not one with which these artists felt happy. The general consensus was that 'conceptual dance' suggests something that is primarily located in the mind, whereas they believe their work is concerned with embodied experiences and with the way these are perceived. This chapter discusses recent dance works by Yvonne Rainer and Xavier Le Roy which use sophisticated conceptual strategies to deconstruct conventional expectations about the nature of theatre dance. Their conceptual nature does not, however, render them purely cerebral. New ways of moving and performing can create new kinds of feelings. By doing so, they can give expression to new ways of perceiving and thinking thus contributing to cultural development and change. Such dance works can thus become political. If they do so, however, this will not be because they convey political messages, or present or celebrate political values or symbols. What they can sometimes do, instead, is bring about a sense of recognition, shared by performers and beholders, of the need for change and of the difficulties that hinder and block those trying to bring about change in complex, contemporary situations. These dance works, I argue, explore the potential of dancing bodies to generate affects that bring people together. What they therefore propose is a kind of libertarian openness that resembles the account of affective sociability proposed by the seventeenth-century philosopher, Baruch Spinoza.

In his *Ethics*, Spinoza pointed out that no-one has hitherto determined what the body can do (2001: Ethics 3, prop. 2, scholium). Our potential as embodied beings, he implies, always exceeds our understanding of our bodies. Spinoza's insistence on the unknown and illimitable potential of the body resonates with the kinds of researches into what the body can do that have been carried out by artists like Le Roy and Rainer. Spinoza's speculations about embodied potential are part of his discussion of the embodied nature of ethics. He believed that ethical behaviour can contribute to the development of just, faithful and honourable patterns of sociability because of the ways in which affect can bring people together. Spinoza's ideas have been taken up in the late twentieth and early twenty-first centuries by philosophers associated with progressive politics including Giorgio Agamben, Judith Butler, Gilles Deleuze, Brian Massumi and Toni Negri. Their application of Spinozan ethics to contemporary political concerns informs my reading of the new kinds of affects which recent dance works by Le Roy and Rainer create. I shall argue that dance pieces, whose performances foreground the process of movement research by rethinking the relationship between individual motor functions, can open up our potential for ethical relations.

There are two separate phases in Rainer's dance work. The first began when she was a member of Judson Dance Theater in the early 1960s and ended when she moved to film-making in the early 1970s. The second began in the late 1990s when she started restaging some of her earlier work, and has seen her make two entirely new pieces, to date, based around music by Igor Stravinsky. One of the key concerns in her choreography has been deconstruction of normative performer audience relations. One strategy she adopted in the late 1960s for prompting dancer and beholder to find new ways of relating to one another was the use of alternative performance modes. These included showing 'work-in-progress', on stage rehearsals, and instances where a dancer learnt existing choreography live on stage. In one of these performances Rainer taught Becky Arnold her best known piece *Trio A*. Her piece *Continuous Project – Altered Daily*, presented in 1970–1, as its title suggests, was an evolving, process oriented piece, and *Grand Union Dreams*, first performed in April 1971, gradually metamorphosed into an ongoing unrehearsed improvisation by the Grand Union who took their name from the piece. Their performances eventually no longer contained any of the material Rainer had initially choreographed. Rainer's recent piece *AG Indexical With A Little Help From H.M.* (2006), which restages material from George Balanchine's 1957 ballet *Agon*, includes a section in which

Sally Silver teaches herself a solo from *Agon* live on stage by watching a recording of it on a video monitor. All of these works put their beholders in the position of observing and appreciating how bodily intelligence can work towards the solution of movement problems. By showing this, the choreography puts dancer and beholder into new kinds of relations with each other.

Xavier Le Roy is one of a group of European dance artists who first began showing work in the early 1990s. He too has sought to blur the distinction between performance and rehearsal. As he stated in his 'Self-Interview': 'I don't accept the dissociation between rehearsals and performance because I think that you cannot separate the representation of the bodies from the set up you used to develop these representations' (2000: n.p.). Le Roy took part in a project to restage Rainer's *Continuous Project – Altered Daily* in 1996. He acknowledges that this experience had a significant impact on his own work during his 1998 autobiographical performative lecture demonstration *Product of Circumstances*. By giving this lecture about himself and inviting the audience at the end to ask him questions about it in the context of dance performance, Le Roy has therefore, like Rainer, explored new modes of performing and new kinds of performer audience relations. As I show, one of his concerns has been to critique how working with and for theatres, arts festivals and other institutionalised agencies can constrain such relations.

Drawing on Spinoza's account of body and affect as interpreted by Gilles Deleuze, this chapter explores correlations between these experimental investigations into what the dancing body can do and Spinoza's vision of ethical sociability. It examines how such dance performances, through their refusal to allow spectators to take up conventional subject positions in relation to dancing bodies, remind audiences that they can never know the full extent of what bodies can do. By challenging beholders to work actively at interpretation, such performances prompt them to find new ways of relating to the performers (and to each other); these, in turn, can suggest new potentials for social and political relations that are underpinned by an openness towards affective experience. This chapter therefore examines the way Spinoza's account of ethics can offer insights into the kinds of social and political relations suggested by Rainer and Le Roy.

Affect and sociability

The term 'affect', initially adopted by psychologists in the 1890s, is often used to refer to feelings in an abstract, impersonal sense. In Deleuze's reading of Spinoza, an affect is a change in energy level which impacts on individual senses qualitatively. What an individual calls joy, for example, is an increase in this energy level, and sadness is a decrease of it. Affects are, therefore, qualities within experiences while emotions are the socially and culturally constructed labels which individuals use when they have identified what their feelings are. By labelling these qualities, they personalise them.[3]

A key concept within Spinoza's account of the embodied nature of affect is his idea of the *conatus*. This is usually defined as the striving to endeavour to persist in being, or the desire to persevere in one's being. One of the surprising aspects of Spinoza's account of ethics is that it consists of the pursuit of self-preservation, and yet, at the same time, Spinoza believed that such ethical behaviour can lead to just, faithful and honourable patterns of sociability. This poses the question of why an individual whose actions are motivated by self-preservation would need to acknowledge the interests of others. In Spinoza's view, individuals not only have the power to influence others, but are also vulnerable to influence from others that may or may not be beneficial. He argues that social interaction involves a flow of affect between individuals, and that: 'The mind, as far as it can, endeavours to conceive those things which increase or help the power of activity in the body' (Spinoza 2001: Ethics 3, prop. 12). Spinoza was in dialogue with Descartes's theoretical distinction between mind and material body. As Genevieve Lloyd notes, 'Spinoza's physics of bodies, and his treatment of the collective powers of socialised bodies, seem to involve both a preservation and a blurring of the limits between bodies' (1996: 95–6). Spinoza argued that it is in people's interest to protect themselves from harmful influences and ensure they only come under beneficial ones. He proposed that if 'two individuals of entirely the same nature are united, they form a combination twice as powerful as either of them singly' (Spinoza 2001: Ethica 4, prop. 18, proof).

In Spinoza's account, all the emotions derive from what he called the 'primitive affects' of desire, joy and sadness. Each of these involves the transition to a greater or lesser power to do things. According to Spinoza, these affects are caused by the presence of the body of another person. He argued that if 'the nature of the external body be similar to the nature of our body, then the idea which we form of the external body will involve a modification of our own body similar to the modification of the external body' (Spinoza 2001: Ethica 3, prop. 27, proof). Spinoza is not talking about material modification of a machine-like body, but momentary modification in the state of a dynamic process of embodiment. If one is inevitably influenced by others, then, as Moira Gatens notes: 'an attempt to organize one's encounters in order to minimize bad and maximize good affects leads human beings to sociability' (Gatens 2000: 65). It is these interactive, relational possibilities, I suggest, that Rainer and Le Roy have sought to explore, in different ways, through their process-oriented dance works.

Kinaesthesia

Questions about the status of affects within artistic production were evidently at the front of Rainer's mind around the time when she was considering shifting the focus of her creative practice from live dance performance to film-making in the early 1970s. In a letter to the art critic Nan Piene, written in 1973, Rainer outlined her reasons for making this transition. The problem for her was as follows: 'Dance *ipso facto* is about me (the so-called kinaesthetic response of the spectator notwithstanding, it only rarely transcends the narcissistic-voyeuristic duality of doer and looker); whereas the area of the emotions must necessarily concern both of us' (1974: 238). In her view, most dance performance positioned beholders in passive, voyeuristic positions – as connoisseurs of the dancer's charismatic, rarefied sensibilities. Rainer wanted to deflate dancers' narcissism and make beholders engage more actively in apprehending the dance. In other words, she envisaged a situation in which dancers and beholders would be more equal so that they could recognise what she calls the area of emotions – but I am calling affects – as something that, through kinaesthesia, brings them together.

Peggy Phelan (1999) and Carrie Lambert (2002) both recognise the centrality of affects within Rainer's work. Lambert proposes that 'Empathetic experience on the part of the spectator is the founding principle of both Rainer's chosen media – dance and film – in their conventional forms. In cinema it is called identification; in dance, kinaesthetic response' (Lambert 2002: 41). The kind of empathy she identifies in Rainer's dances and films is one that depends upon beholders taking an active, imaginative step towards the affects produced during the dance performance. Rainer's presentation of ordinary movement as dance, Lambert suggests, encourages beholders to empathise with the actual muscular feelings and physical sensations that the dancer has while performing it. It is this understanding and sharing of the physical sensations of motor activity that Lambert calls kinaesthetic empathy. In her view this is, in effect, a limited kind of empathy and one that is further circumscribed because of the effects of voyeurism and narcissism. Lambert, therefore, suggests that Rainer moved to film-making in order to be able to introduce a wider range of materials and subject matter into her work. Lambert's conclusion is that Rainer's films offer the spectator the pleasures of 'identifying with and kinesthetically bonding with characters', providing that they 'pay attention and are willing to projectively imagine' (2002: 63).

The theoretical premises on which the idea of kinaesthetic empathy are based have been criticised by dance scholars. The term 'kinaesthetic' is of comparatively recent origin. As Hanna Järvinen has pointed out, when early nineteenth-century neurologists initially proposed that sensory perception was entirely separate from motor function, they effectively excluded any possibility, in scientific terms, of proposing that the activity of dancing could be a sensory experience. The English neurologist Henry Charlton Bastian invented the term 'kinaesthesia' in 1880 to account for his observations concerning the source of sensory information about motor activity. Järvinen suggests that the term has been taken up, particularly within dance theory, as if it were a separate sense in addition to, but separate from and transcending, the existing five senses:

> Rarefied kinesthesia as the 'dance sense' is usually employed to create a site of privileged knowledge where dance is primarily aesthetic experience by the dancer (or, in the case of the theories of metakinesis, kinesthetic empathy and kinesthetic sympathy, a mystical connection between the body of the dancer and that of the spectator-expert).
>
> (2007: 147–8)

It is therefore necessary to detach the term 'kinaesthesia' from some of its past associations in order to understand how different senses combine and interrelate when we experience the kinds of affective qualities that are produced during dance performances.

While Peggy Phelan also discusses Rainer's work in terms of empathy and kinaesthesia, she gives a much more complex account than Lambert of the way Rainer's dances and films produce social and political meanings, particularly in relation to feminism. Phelan concludes her discussion of the relationship between Rainer's work in dance and film by proposing that:

> In 1965, Rainer yelled 'no to moving and being moved' but the rest of her effort has been geared towards finding the generative intelligence

> produced by, and necessary for, kinaesthetic, political, and psychic empathy. A well-toned empathy increases our capacity to respond to the compelling possibilities of moving and being moved and remains our best hope for remaking ourselves and our world.
>
> (1999: 16)

Whereas Lambert sees the kinaesthetic as a means through which specific muscular feelings and physical sensations are communicated, Phelan seems to understand it as a process for sharing feelings about social and political experiences. These are not, however, emotions that are felt by the performer and then externalized. Phelan implies that they are produced by the connections created between the art work, its makers and performers, and its beholders. Phelan is interested in what happens when the work engineers a coming together of these parties.

The terms kinaesthesia and empathy imply social and communal relations, but, as Järvinen notes, these terms can get in the way of understanding the processes they name if they posit dance as a unique essence that transcends ordinary experience. Recent scientific research suggests that sensory perception exploits the interconnectedness of what have previously been considered separate faculties (see Bleeker 2006). Phelan's vision of people working together to make a better future is one that resonates with much left-wing political thinking. What is distinctive is Phelan's understanding of the role that affects, generated in Rainer's work, can play in bringing people together. What she describes corresponds in many ways with Spinoza's belief that ethical behaviour can lead to just, faithful and honourable patterns of sociability. As I noted earlier, in Spinoza's view, individuals not only have the power to influence others, but are also vulnerable to influence from others that may or may not be beneficial. In Rainer's binary of voyeurism and narcissism, individuals fail to interact in a beneficial way. It is through opening up in a positive way towards affective qualities that a potential for new kinds of shared experiences and new ways of relating with others become perceptible.

AG Indexical With A Little Help From H.M.

As I noted earlier, there were a few instances during the late 1960s and early 1970s when Rainer explicitly taught dancers some of her choreography live on stage during a live performance. She returned to this idea in her 2006 piece *AG Indexical With A Little Help From H.M.* – henceforth *AG Indexical*. During this, Sally Silver copied a male solo from

Balanchine's *Agon* (1957) by watching a video recording of it. As Rainer explained to Clarinda Mac Low:

> I didn't want her to learn it, but rather wanted her to do it spontaneously while looking at the monitor. So, what she does [...] she turns the monitor around so she sees it, the audience doesn't. She will be watching it and trying to do the steps. So not only is her learning process being presented as performance, but the obvious fact that Sally is untrained in ballet, and as a woman is doing highly-trained, skilled, male ballet dancer's stuff.
>
> (Mac Low 2006: n.p.)

After seeing this once in rehearsal Rainer asked Silver not to work on it any more so that the spontaneity would still be there in performance. When I watched this section in the theatre, the audience applauded Silver. Knowing what Silver was doing, they seemed touched by it, as if she was no longer just a dancer on stage but was revealing that she was a human being as well.

Since Silver was herself learning the solo, she was using the filter of her own perception, in real time, to select from the dance information that the video offered to her. While doing this, she was using her own memory and experience to assemble and refine her re-presentation of the movements that make up this solo from *Agon*. As Brian Massumi suggests, 'Perception is never only impression. It is already composite. Studding each impression are shards of intentions and conscious memories' (2002: 75). What beholders could therefore apprehend was not a pure essence of this solo from *Agon*, but its choreography together with the qualities that Silver herself inevitably added to it. These came from both Silver's and their own memories from past experiences together with traces of her creative, embodied intelligence. Rainer's friend and colleague Trisha Brown once spoke about the unpredictability of dance. The body, she said, 'doesn't move with the clarity of line or mechanics that I wish for. Dancing is like scribbling, you know, because of the inconsistencies of human anatomy. [...] It's the human failure factor in the exposition of form that makes for this marvellous thing called dance, which is highly imperfect from the beginning' (in Goldberg 1991: 6). What Silver added to Balanchine's solo by performing it in this way was the human failure factor that Brown says makes this marvellous thing called dance. While one could see this as a quality uniquely individual to Silver, what was marvellous about it was not its uniqueness, but the fact that it was emblematic of an immanent human potential.

Inevitably, *AG Indexical* was particularly challenging for beholders who normally only watch classical ballet. John Rockwell in the *New York Times*, for example, complained that the piece 'looked like a not-too-amusing parody of ballet' and that despite the personalities and importance of the dancers, they 'looked like amateur parodists' (2006: n.p.). What is significant about this review is the underlying critical stance Rockwell takes up. He judges the reworking which Rainer and her dancers created against Balanchine's canonical position as someone whose work expresses a universal essence. *Agon,* in this view, is complete in itself and not something that warrants any supplementary signifying layers, especially if these are inspired by feminism. To judge *AG Indexical* against the standard of Balanchine's original performance is to situate performers and beholders in a hierarchical worldview constructed through judgements of supposedly universal values.

The demand of universality constrains dancers to subordinate themselves to an external standard of critical judgement. A dance work, however, that refuses to comply with these hierarchical demands, creates a situation in which the performer is able to investigate, in her own terms, the question of what the dancing body can do. At stake is the difference between a constraining system of critical judgment and a libertarian openness towards embodied potential. The latter, I claim, corresponds with Spinoza's account of ethical sociability. The distinction between hierarchical judgement and libertarian ethics which I am making here is one that Deleuze identified during a seminar on Spinoza. Deleuze argued that morality is a system of judgement, and that judging always implies an authority external to being. Ethics, he argued, is not about judgement but is instead anti-juridical. It is about the question of what, for an individual, it is and is not possible to do. As he explained:

> In a certain manner you say: whatever you do, you will only ever have what you deserve. Somebody says or does something, you do not relate it to values. You ask yourself how is that possible? How is this possible in an internal way [...] The point of view of an ethics is: of what are you capable, what can you do? Hence a return to the sort of cry of Spinoza's: What can a body do?
>
> (Deleuze 1980: 3)

Deleuze's interpretation of Spinoza is one that stresses potentials that are anti-juridical and egalitarian. Someone who thinks the dancers in *AG Indexical* looked like amateur parodists cannot recognise the new potential which the work opens up and displays particularly if he fears the feminist egalitarianism informing it.

Just as Rainer valued Silver's potential for dancing material conventionally coded as male, she was also fascinated by the way the male dancer in *Agon*'s final duet manipulates his female partner. In Rainer's reworking, three experienced, female contemporary dancers – Silver, Pat Catterson and Patricia Hoffbauer – help Emily Coates, a younger ballet-trained dancer, to take up the difficult sequence of ballet positions that Balanchine had choreographed for the ballerina. Rainer had read that Balanchine choreographed *Agon* shortly after his then wife, Tanaquil Le Clercq, had been paralysed with polio. Balanchine had worked hard with her on physical rehabilitation, manipulating her paralysed legs into ballet positions. Rainer applied this image quite literally so that Coates 'hardly does anything on her own. Her every movement is initiated by someone lifting her legs or hauling her about' (Mac Low 2006: n.p.). If three men had manipulated her in this way, the effect would have been disturbing. Three women helping a fourth to achieve such extreme flexibility suggests positive values of caring and co-operation. It thus performatively affirms an aspiration towards ethical behaviour that can lead to better patterns of sociability.

Product of Circumstances

In his lecture demonstration *Product of Circumstances*, Xavier Le Roy says that his participation in the project to restage Rainer's *Continuous Project – Altered Daily* (*CP-AD*) not only gave him a greater understanding of history but also suggested solutions to questions he had about how to carry out research into what the body can do. Rainer's *CP-AD* confirmed for Le Roy the possibility of creating choreography out of material that did not conform to conventional vocabularies of aesthetically valorised dance movements. The discovery of ordinary, task-based and pedestrian movement in Rainer's work affirmed Le Roy's own research into similar kinds of movement. The emphasis on process in Rainer's piece offered Le Roy an answer to the question: 'Can the production of a dance piece become the process and the production in itself without becoming a product in terms of performance and representation?' (Le Roy 1998: n.p.). It is this question that he set out to address in *Product of Circumstances*. In this piece Le Roy talked about his postgraduate research in molecular biology, and then his gradual development of a career as a dancer. He illustrated it with slides of photographs and diagrams from his PhD thesis and extracts from dances he had choreographed. The extracts included sections from earlier pieces *Narcissus Flip* (1997) and *Self Unfinished* (1998), which investigate what Le Roy calls the functions and dysfunctions of the body. These show the process he devised for analysing dance movement by taking apart the dancing body and creating links between its fragments. This sometimes resulted in strange, seemingly inhuman movement sequences. One of these explores isolated movements of the lower arm swinging from the elbow. Le Roy stood on a chair facing the audience and swung his lower arm in different ways while keeping the rest of his body still. Mime uses isolation to break movements down into smaller and smaller units so as to create language-like segments out of which to build a narrative. Le Roy's movements, however, did not create a narrative. They were movement experiments that explored the potential of isolated elbow and lower-arm movements for generating affects to discover if any of these had not previously been considered as dance.

Like *AG Indexical, Product of Circumstances* is, in Deleuze's terms, concerned with ethics rather than judgment, and indeed draws attention to the inappropriateness of making judgments. Discussing his early experiences learning dance in the late 1980s and early 1990s, Le Roy showed the length of his arms in relation to his height and, bending over as if to touch his toes, demonstrated that his back was not very flexible. He performed some Cunningham-esque class exercises, but explained that he had had difficulty attaining the kind of polished mode of performing that is generally looked for during an audition for a mainstream dance company. *Product of Circumstances* presented these as findings from his research, just as, elsewhere in the piece, he projected microscope slides of human breast tissue that had been treated with different chemicals, and diagrams analysing the data derived from them. For Le Roy, molecular biology and theatre dance are fields of research into what the body can do.

Le Roy explained that while writing his PhD, he became disillusioned with institutional demands to publish and raise grant money. 'I was learning that research has to follow the methods of capitalism. I was asked to produce science not to search' (1998). This is why, after completing his doctorate, he abandoned scientific research and began to work in dance. However, when he began to receive invitations to present dance work, he realised that:

> the systems for dance production had created a format which influenced and sometimes to a large degree also determined how a dance

> piece should be. Most of the time producers and programmers have to significantly follow the rules of global economy.
>
> (Le Roy 1998)

In other words, he became aware of similar institutional pressures to make marketable products: 'I felt like a fugitive who actually never escaped' (ibid.). He was still being judged by a supposedly universal, external standard of judgment.

While the process of research informing works like *Narcissus Flip* grew out of his increasing disappointment with mainstream contemporary dance, *Product of Circumstances* has taken this in a more explicitly political direction. Its deconstruction of normative aesthetic ideologies is, in effect, a strategy for resisting the embodied effects of conventional power relations institutionalised in the dance world. Through his emphasis on process rather than product, his work evades conventional aesthetic identification and, by doing so, creates a space for a biopolitics in which the dancing body becomes a site of resistance. As such it performatively suggests the kind of evasion of normative social and political identifications that the Italian philosopher Giorgio Agamben has recently discussed. Agamben suggests that modern nation-states

> are not founded on a social bond, of which [the state] is the expression, but rather on the dissolution, the unbinding that [the bond] prohibits. For the state, therefore, what is important is never the singularity as such, but only its inclusion in some identity, whatever identity (but the possibility of the *whatever* itself being taken up without an identity is a threat the state cannot come to terms with).
>
> (Agamben 1993: 85, his emphasis)

Le Roy's relationship to the dance world resembles the relationship Agamben proposes citizens have with the state. Agamben's anti-juridical view of human potential resonates with Deleuze's interpretation of Spinoza. The word 'whatever' is crucial here. Whatever (*qualunque* in Italian) does not mean, in a negative sense, 'it does not matter which, indifferently', but positively 'being such that it always matters' (ibid.: 1) and a 'singularity exposed as such [that] is whatever you want' (ibid.: 2). In Agamben's terms, Le Roy's work creates whatever aesthetic identity which arises out of a political intention to redefine dance as whatever the dancing body can do, as dancing-in-process rather than choreography-as-product. As such it would seem to be in conflict with the globalised economy of the dance world.

Conclusion

I have uncovered historical links and similarities of intention between Rainer's presentation of rehearsals as performances and Le Roy's emphasis on process rather than product. Rainer's and Le Roy's investigations of what the dancing body can do can be seen as ethical practices. In Spinozan terms, the affective qualities produced in their dance works performatively propose practices of the self that have the potential to lead to more open patterns of sociability. Rainer's rehearsals celebrate and affirm individual potential for liberation from hegemonic norms. Her *AG Indexical* also articulates a feminist aspiration for collective transformation. Le Roy's work focuses its critique on the way the dance world replicates the workings of state power and global economy. His researches into what the dancing body can do have uncovered bio-political strategies for resisting normative, institutionalised power relations. Where Rainer's group pieces investigate kinds of social collectivity that form through affirmation of shared aspirations, Le Roy's work troubles institutions by evading conventional identification. What they both share is the belief that the search for new ways of moving can reveal potentials for imagining future utopian alternatives to existing social relations.

Notes

1. An early version of this chapter was presented at the first Encontro Internacional De Dança E Filosofia: O Que Pode A Dança?, organised by The Federal University of Rio de Janeiro and the Centro Universitário da Cidade – UniverCidade (2005). I am grateful to Charles Feitosa, Roberto Pereira, Thereza Rocha and Dee Reynolds for their comments.
2. The panel discussion titled 'Not Conceptual' (22nd February 2007) was part of the series 'Parallel Voices' at the Siobhan Davies Studios. It is available as a podcast at http://www.siobhandavies.com/index.php/parent/68/item/474
3. For a discussion of the difference between affect and emotion in 1930s US modern dance, see Franko 2002: 38–58.

Section 7
Conclusion

Introduction

As we explained in the general introduction to this volume, one of our aims in bringing together the chapters for this book has been to reflect on how and why our shared agendas have formed and evolved. So far we have touched on this briefly in the introductions we have written to individual sections, and we address this in more depth in this final section of the book. In 1998, at the inaugural meeting of the *Corporeality and Choreography Working Group* of the *International Federation for Theatre Research (FIRT/IFTR)* we presented a jointly written 'Position Paper'. In this we set out what were the key issues and ideas that, in our opinion, our field needed to address. Since 1998, either singly or together, we have attended all but one of this working group's annual meetings. Several of the chapters in this book have developed from papers presented for discussion during these meetings. In many ways our 1998 'Position Paper' sketched out most of the political concerns and methodological approaches that inform the chapters in this book. Reading it again, ten years later (at the time of writing the current piece), we have also been able to reflect on the extent to which these concerns and approaches have developed and changed. This final section is therefore structured as follows. First, it contains our 'Position Paper', published here for the first time without any rewriting (but without the two personal reflective statements that accompanied it). Following this, we discuss the ways in which this paper anticipated the concerns and approaches developed in the chapters that make up this book while also noting where our agendas have changed. We then locate our own partly shared intellectual journeys within the development of Anglo-American Dance Studies, presenting some critical reflections on its development and its current state. Using this as a springboard we then set out what we, in all humility, believe the key issues and ideas are that, in our opinion, our field needs to address in the future. The section concludes with reflections on the 'intellectual friendship' that has produced this book and its recurring concern with openness, affective relationality and ethics, which occur both within the examples discussed and within our collaborative research processes. It thus reviews what it has meant to us to be writing dancing together.

FIRT/IFTR Position Paper – 1998

Investigations towards the theorisation of embodied representations of gender, 'race' and sexuality in dance performance

AIM

Our aim is to develop new methodologies and epistemologies for undertaking theoretical, historiographical and performative research into the ways in which the choreography and performance of theatre dance represents aspects of identity, relating in particular to sexuality, gender, 'race', and national identity or internationalism. These frameworks for research will be underpinned by new methodologies and epistemologies for conceptualising and analysing the relationship between the dancing body and embodied subjectivity, drawing on areas that are cognate to dance scholarship, in particular those building on French poststructuralist and postmodern theory, feminist theory and queer studies.

INDICATIVE CONTENT

Dualism: residual Christian/Hellenic ideas about the body

Discussion:

(1) It is often said that the body is the primary means of expression in dance. This is, however, based on a dualistic conception of embodiment where:

- the body: mind/body, soul is thought of as ghost in the machine
- inside/outside: that there is always already something inside which can then be expressed (ex-press = push outside)

Furthermore the body remains unvalued within an underlying hierarchy in Platonic and Aristotelian philosophy where 'the word' is primary and originary, while the bodily experiences of the individual subject lead to corruption of the purity of ideal, intellectual comprehension. Within Aristotle's conception of mimesis, the arts only imitate, and bodily imitation is inferior to poetic imitation as only the latter reminds the soul of higher truths that are already inscribed on it.

Catherine Bynum (1995) has drawn attention to the commonplace misapprehension that in Western society from Plato to Descartes the body was universally conceived in dualistic terms of body and spirit. Many medieval thinkers, she points out, held monist or trinary views of the body, some of which she even compares with the antifoundationalist ideas of Judith Butler. Descartes nevertheless, as Susan Bordo observes, conceptualised the body 'as the site of epistemological limitation, as that which fixes the knower in time and space and therefore situates and relativizes perception and thought' (1995: 227); in this view of embodiment, the thinking subject needs to transcend the body if s/he is to achieve rational objectivity.

(2) Any account of theatre dance and representation needs to take into account the relationship between language and the fact that subjectivity is embodied

There is a need to deconstruct the binaries that underpin conventional dance theory, because they limit our conceptualisation of a much more complex process. They constrain our understanding of representation in dance because one side of each binary is valued over the other, producing power relations which result in discriminatory practices.

In the following list the first concept is conventionally valued over the second:

- outside/ inside
- mind/body
- same/Other
- masculine/feminine
- white/black
- Western/non-Western
- heterosexual/homosexual
- 'natural' body/socially constructed body
- form/content
- signifier/signified
- symbolic/imaginary

There is a need to rethink theatre dance and representation by considering the space between these binaries, thinking of it as a range of differences or potentialities or what Deleuze and Guattari call a 'plane of consistency' and to focus on what Julia Kristeva has called the hinge where subject and object flow together.

Representation and theatre dance

Since Descartes, the subject has been thought of as situated in and positioning her/himself in space and time; and poststructuralists have added that the subject is also located within discourses. Choreography and performance

→

as representational practices construct embodied subjectivity, and conventionally position the spectator as subject and the performer as object in particular relationships to time, space and discourse. Theatre dance therefore constructs subjectivities:

- in space – around the body, through space, through combinations of bodies in space, through socially constructed spaces, through ways of understanding space
- in time, in different experiences of time
- in discourses and ideologies

It is our contention that, while space and time are frequently seen as unproblematic in dance theory, they are central to the processes of representation. Representation itself is under-theorised within dance theory. Most theorisation of theatre dance has been concerned with aesthetic appreciation and posited upon an idea of subjectivity that has its roots in the Enlightenment, in particular in Kant's view of critical judgement. While the idea that aesthetic value is 'natural', timeless and universal does not altogether rule out representation, the notion of the rational unitary subject of the Enlightenment constitutes an account of the embodiment of subjectivity in time and space which has implications for the theorisation of dance as a representational practice, particularly through the development of the notion of the proscenium arch as a fourth wall, which is discussed below. Poststructuralists (among others) have produced a critique of the rational unitary subject which is based on acknowledgement of the materiality of the body. Recent attempts to theorise the body offer new ways of theorising choreography and performance.

As John Forrester has observed, the fundamental attack on the necessarily unified character of the subject has come from psychoanalysis. While recognising the importance of ideas about the subconscious in contemporary cultural analysis, psychoanalytic theory is not the focus of our investigations. While we accept the Freudian idea that the ego is subjected to the force of desire 'of which it knows nothing, or of which it only hears the most distant and distorted reports' (Forrester 1987: 15), this raises the problem of how to conceptualise desire, and how to interpret these 'distant and distorted reports'. As Elisabeth Grosz points out, from Plato to Freud and Lacan, desire is predicated upon lack, and on a 'yearning for what is lost, absent, or impossible' (1994: 120) which is projected outside the subject's body on to the body of another whose difference makes them Other. These theories have implications for both gender and sexuality because the lack is theorised as lack of the phallus. Queer theorists have argued that the idea that desire is predicated upon lack enforces normalising ideologies of heterosexual sexuality. Psychoanalytic accounts of representation in cultural forms which are posited on the notion of desire as lack are therefore problematic. Nevertheless much poststructuralist theory incorporates elements of psychoanalytic theory, drawing particularly on the work of Lacan. We therefore choose to remain sceptical about psychoanalytic theory and only intend to use aspects of it where necessary and in a pragmatic and provisional way.

Notions of subjectivity, discourse, space and time

Discussion:

(1) The embodiment of the rational unitary subject

The philosophical notion of the rational unitary subject brackets off aspects of embodied experience that are deemed a hindrance to the making of rational, disinterested critical judgements, including aesthetic judgements. Where theatre dance is concerned, it is significant that the rational unitary subject as embodied subject is reduced to finite co-ordinates in time and space, where time and space are seen as quantifiable and measurable in a scientific and mathematical way. By bracketing off other ways of experiencing time and space, the subject is confined to a private time/space whose isolation from the private time/spaces of others ensures the disinterestedness of individual judgement. These new, rational ideas of the rational unitary subject as embodied subject in time and space underpin ideas about the spectacular nature of the presentation of theatre dance as this developed in the eighteenth century. Thus as the rational unitary subject was confined to the measurable temporal and spatial co-ordinates of a body that had become an object of verifiable, and hence repeatable, scientific knowledge, so too the dancer's movements became verifiable elements within a lexicon of schematised steps whose repeatability was guaranteed by the method of Feuillet notation.

In a discussion of eighteenth-century performance theory, Joseph Roach (1985) posits a connection between medical models of the body and methods of acting. Developing a Foucauldian account, he argues that Graeco-Roman ideas of acting and oratory (which continued to be influential up until the mid-eighteenth century) depended on a medical idea of the body where the actor or orator literally excited his or her audience through bringing about a physical effect on the listener's/spectator's body. Thus the dancer or actor sought to involve her or his patrons (some of whom in London theatres may even have sat up on stage at the sides of the performance space) in an intimate and engaging performance. New naturalistic ideas about acting which developed during the eighteenth century accompanied an increasingly modern, 'scientific' understanding of the anatomical functioning of the body. As the body became the object of what Foucault calls new regimes of power knowledge, theatrical reformers transformed plays and ballets into a spectacle of carefully constructed moving tableaux. Not only were audience members banned from sitting on the stage, but they were further separated from the performers through the idea that the proscenium arch was an invisible fourth wall. No longer seduced or harangued by performers, the individual audience member was now free to exercise her or his critical faculties in a disinterested manner. As the rational unitary subject became subject to regimes of power knowledge, so the actor's presence was reduced to a visual hieroglyph in a moving tableau.

→

(2) Embodiment and the postmodern subject

Representation is a two-way process involving individual agency and inscriptions through socially constructed discourses of normative social values. As Roland Barthes points out, individuals are located within normalising discourses: 'I am obliged to posit myself as subject before stating the action which will henceforth be no more than my attribute. [...] To speak, and, with greater reason, to utter a discourse is not, as is too often repeated, to communicate; it is to subjugate' (1983: 460).

This two-way process means there is a continual vying for power between a sense of individual agency which is part of 'subjectivity' and the ways in which 'subjects' are located in, and inscribed by, discourse. This means that subjectivity is in a continuous state of flux. There is therefore no 'real' in the sense of a transcendent Platonic idea inscribed on the soul to which representations in dance refer in a unidirectional way. Poststructuralist theory recognises no original which can stand as guarantee of the authenticity and correctness of the representation. Some poststructuralist and queer theorists have taken up a deconstructive strategy of reversing the direction of this process. Thus, for example, whereas the rational unitary subject constructs the social world out of the objects which s/he becomes aware of through her/his senses, Baudrillard has adopted the 'fatal strategy' of trying to think the social world from the point of view of the object. Judith Butler has adopted a similar device arguing that gendered identity is not expressed through performance, but formed through performing acts which constitute it. While Butler's notion of the performative nature of gender should not be conflated with the performative nature of theatrical dance, the performance of theatre dance does reiterate and reinforce the construction of gender in the way that Butler has theorised: 'producing the effect of an internal core or substance [...] on the surface of the body, thorough the play of signifying absences that suggest, but never reveal, the organising principle of identity as cause' (Butler 1990: 136). In Butler's view, the subject constitutes her/himself through self-constituting acts which s/he is compelled to repeat. This performance involves repetition but it is always unfaithful repetition both based on and producing misrecognition. This unfaithful repetition has been theorised elsewhere in varying ways as mimesis or mimicry (Irigaray cited in Butler, 1993; and also Deleuze and Guattari, 1988), and as an echo (Kierkegaard cited in Battersby, 1998) and as a 'riff' or 'ritournelle' (Deleuze cited in Battersby, 1998).

In this performance, which involves a play of signifiers and which exists in space and time, a relationship is forged between the performer and the spectator which is continually changing. For example, since we no longer accept a notion of a rational unitary subject, the distance between performer and spectator has changed, there is no longer a 'fourth wall'. This has implications for notions of space and time. If space is non-hierarchical and not framed or limited, then notions of 'in between' spaces, or borders or frontiers become possible where more than one genre of discourse co-exist, thus resulting in hybrids. This space cannot be measured in the traditional sense (mathematically). Opening up and reconfiguring space in this way allows for the possibility of new narratives which could represent embodied subjects.

Similarly time is non-linear. As Deleuze, following Bergson, has pointed out, there are other ways of experiencing time than those which can be framed or measured in the traditional mathematical sense. Kristeva, in a much cited essay, has raised the problem of the gendering of monumental, cyclical experiences of time as female in contrast to the notion of time as linear, cursive history. Notions of time also need to be opened up and reconfigured.

Underpinning questions of how the subject is located in both time and space are relations of power. To reconfigure space and time is to allow for the possibility of seeing/conceiving more than one thing at a time in a non-linear, non-hierarchical way, and hence for the possibility of seeing new kinds of relationships between things. In other words (literally) language and what Lyotard calls genres of discourse are also reconfigured. The relations between things in representation – the différance/différence/différend (Lyotard) are the keys to how representation, language and discourse operate. We are concerned with theorising representation in theatre dance as an instance of what Barthes calls a 'writerly text', where the interpretation of the text is opened up and the relationship between writer and reader/performer and spectator is deconstructed. This allows for communication or reading in the gaps and excesses of performance (Phelan 1993) which we are concerned to investigate. Through identifying these gaps and excesses it becomes possible to see/read and write/perform the unmarked subject – ungendered, unsexualised, unracialised – thus dismantling the binaries by considering the space in between them.

By investigating representation in theatre dance as a two-way process which involves relations between individual agency and a situated (in discourse) subject mapped on/traced over/in dialogue with a three-dimensional process involving the body in performance in space and time, it becomes possible to imagine reconfigured bodies in reconfigured space and time and thus enable the performing/writing/reading of new types of narratives. These bodies and narratives exist in a three dimensional matrix that is continually changing:

SPACE/SPACE

BODY/BODY

TIME/TIME

Within each of these dimensions the two-way battle between agency and social construction/situation is continually being forged, hence two of each – this is because, on one level, the subject is positioned within and outside discourse, and, on another level, unfaithful repetition occurs involving the continuous plays of signifiers. We are therefore concerned with representation as a process of becoming rather than being, and with understanding how representation works rather than with establishing what representations mean.

Further reflections, 2008

One of the key changes which underpins the ways in which we have developed our ideas since writing this in 1998 has been our move from a concern with binaries to a focus on multiplicities. The importance of this approach is underlined and explained in Valerie's discussion of *Shiver Rococo* in Chapter 3, and this is taken up and explored further in other chapters, such as our discussion of *Yippeee!!!* in Chapter 6. Implicated in this focus on multiplicities is the importance of openness in several senses. We anticipated this in our 'Position Paper' when we recognised the need to reconfigure space and time in a non-linear sense. Rethinking temporality before spatiality, which is introduced in Valerie's discussion of *Shiver Rococo*, is developed further in both our chapters in Section 5, 'Insurmountable Memories. Furthermore the ways in which an emphasis on openness allows for reading in the gaps and excesses of performance, introduced in our 'Position Paper', are taken up and developed in, for example, Ramsay's discussions of *Napoli* and *Palermo, Palermo* in Chapter 5 and of *Ellis Island* in Chapter 10. Openness is evident in another sense in our recognition of the importance of unfaithful repetition, also anticipated in our 'Position Paper', which is developed in our discussion of *Yippeee!!!* and both our chapters in Section 6, entitled 'Reimagining Dancing Together'. The idea of unfaithful repetition in turn troubles the normative performer/beholder relationship which, as we recognised in our 1998 paper, needed deconstructing. Ramsay's chapters discussing *L'Apres-midi d'un faune* and *Façade* and *Elvis Legs* are examples where the need for a new role for the beholder is addressed. Although all of the chapters here deal with the agenda we set ourselves in 1998, they also demonstrate the extent to which our concerns and approaches have developed and changed.

In our 'Position Paper' we expressed our ambivalence about employing psychoanalytic theories in cultural analysis, recognising their usefulness while remaining sceptical about their ahistorical and normative tendencies. Indeed, this ambivalence was already stated by Valerie in her chapter 'Dancing Dicks', written originally five or so years earlier and reproduced here. What is evident in 'Dancing Dicks', however, is a focus on representations which is also present in our 'Position Paper'. This is no longer as important or central to our work a decade on. One key change for us has been a shift away from a concern with what the dancing body represents. We now recognise the need to develop ways of articulating the experiences of dancing. This involves moving from a notion of representation as a fixed image that is conceived spatially in order to prioritise the importance of temporal experience. Most of the chapters in this book reveal the presence of ongoing, developing, affective intensities. Where these chapters differ from 'Dancing Dicks' is in their conceptualisation of desire. Whereas psychoanalysis theorises desire as lack that is satisfied by another marked as different and Other, these essays now explore the potential of desire to produce multiple intensities, energies and flows. These are all qualities that we experience when beholding dancing.

The other thing that was clear in our 'Position Paper' was a commitment to investigating interconnections between dance and politics. It concluded with a recognition of the importance for this of understanding how representations work, because this created the conditions of possibility for redressing the balance of unequal power relations. Ten years on, there has been a shift from examining what a work represents to a recognition of what performatively it can do. Recognising that a work of art rarely has the power to redress unequal power relations, we are now more concerned with its potential for finding ways of imagining what new, more equal ways of ethical coexistence might be and how these might lead to what is sometimes called the politics to come. Ways of thinking about ethics as a practice of the self enable two interrelated developments. On the one hand, they lead to new ways of thinking embodied experience that no longer separate mind and body but focus on what it is possible to do, the potential for action. On the other hand, these practices of the self enable the emergence of micro-political alliances. These are embodied practices which are foregrounded in the issues and concerns of the particular dances we write about. For instance, Ramsay argues that Isaac Julien's *Western Union* enables the beholder to recognise migrants as people of a coming future. Similarly in Valerie's discussion of Bausch's work, Endicott, Großmann and Shanahan all open up new possibilities for existence by acknowledging its vulnerable and irreparable nature. What has changed since 1998 is a recognition of how much more entrenched and multiplicitous power relations are. Hence there is a need to develop more open ways of thinking that enable us to recognise these multiplicities, particularly within the relationship between dance and politics.

We suggested in the introduction that the political concerns which have remained a consistent undercurrent within our writing now seem more difficult to write about than they were in the 1990s. This is partly a consequence of the demise of the left as a mass movement – with the decline of grass-roots socialism and the collapse of Soviet communism. It is also a consequence of the way dance scholarship has developed in the United Kingdom and United States. Writing in 2006, Susan Manning suggested that a split developed in the mid-1980s between US dance scholars over the development by some scholars of 'methods of cultural analysis derived from anthropology' (Manning 2006b: 1). She argued that a subsequent 'theoretical turn' in dance history has resulted in a shift in terminology so that the term 'dance studies' has gradually replaced the term 'dance history' as a way of designating the field. Starting around 2000, she suggests, 'many commentators regularly opposed the "old" dance history with the "new" dance studies' (Manning 2006a: 2). This 'theoretical turn' in dance studies was largely brought about by scholars who wished to investigate the ways in which normative ideologies of gender and ethnicity could be reinforced through the conventions and traditions of ballet and contemporary dance. In our opinion, controversy surrounding how to discuss the way ballet in the twentieth century has come to represent gender, 'race' and sexuality has led to polarization between the formalist approaches to aesthetic issues used within the so-called 'old' dance history and sociologically oriented approaches to reading dance within the 'new' dance studies.

One consequence of this polarisation has been a lack of critical debate about the status of the dance work within dance scholarship. The French scholar Isabelle Ginot has recently pointed to a tendency within what Manning calls the 'new' dance studies, 'to use choreographic works to demonstrate or illustrate' particular cultural theories so that the work becomes paradigmatic of social and cultural relations rather than being recognised as a work in its own right (Ginot 2007: 253). The way in which a work is read, she suggests, is too often 'organized around an axis of the recognition of signs and the locating of their more or less conformist or transgressive nature with respect to dominant norms' (ibid.: 256). It is Ginot's contention that French scholars have developed significantly different methods of dance analysis to those of their Anglo-American colleagues and her essay calls for more dialogue between these different traditions. Her particular concern is with the dangers of reductiveness within Anglo-American dance studies. In terms of methodology, Ginot warns against approaches that 'confer primacy to predominantly visible and discursive signs of identity' (ibid.: 253) at the expense of an investigation of qualitative parameters within dancers' corporeal practices. At the level of epistemology, she argues that, rather than treating a work as if it 'contains a stable meaning, pre-codified by culture' (ibid.: 256), it is necessary to consider 'the meaning as unstable, produced by the exchange between the dancer and spectator, which comes to modulate, or to transform the stability of the codes put into play by the choreography' (ibid.: 259).

Developments in recent UK and US dance scholarship imply some awareness of the problems and issues that Ginot has brought together and articulated. Where our own work is concerned, Ginot's criticisms could be legitimately applied to some of what we advocated in our 1998 'Position Paper'. There are, however, correlations between Ginot's position and the ways in which we have subsequently reframed our concerns. Like Ginot, we are concerned with developing methods of analysis that recognise the open-ended, unstable, processual nature of dance works. Rather than applying preconceived methodologies to a work, we have argued that works themselves often necessitate or even demand that beholders take up new kinds of positions in relation to them and that this in turn requires the development of new methodologies. Like Ginot, we wish to avoid methodological approaches that close down meanings, and several chapters of this book include criticism of such reductive closures in the way other people have written about dance. These principles form the springboard for what we suggest are the key issues and ideas that our field needs to address in the future. We present these as three aims:

1 To think of dance movement as something that is always in flux

It is always possible to go further in acknowledging the unfixability of dance movement and thus to recognise within it new potentials for emergent behaviour. What new kinds of relations arise when we look at dance movement in this way and what micro-politics might these relations suggest?

2 To resist the formation of exclusive canons within dance history

Whether we like it or not, the canon now exists. Non-linear ways of conceptualising temporality nevertheless offer alternative ways of constructing genealogies of dance traditions and practices. There are dangers of letting oneself be located by the canon in an exclusive position in relation to the past, but we can actively situate ourselves within an inclusive understanding of the past by constructing our own pathways through the archive of dance history. A project that aims to be inclusive has, by implication, political dimensions.

3 To reconsider the status of expression within dance scholarship

Theoretical approaches that draw on recent developments in the neurological and cognitive sciences can deepen our understanding of our affective responses to the embodied practices of dancing. A shift from conceptualising expression in terms of inside and outside towards thinking in terms of the production of affects has consequences for our understanding both of dancers' practices and of how beholders perceive dance. A sensitivity to the kinds of communities brought into being during the moment of live performance itself may reveal the potential for imagining new kinds of relations that, in turn, have ethical and political implications.

Intellectual friendship

In order to understand our intellectual friendship we need to now situate ourselves more personally in relation to the development of our discipline. Initially in the late 1970s and early 1980s, the British dance scene looked exclusively to the United States. What was formative for the two of us at the time was contact with Deborah Jowitt and Sally Banes, who we first encountered through dance writing workshops they taught in London. Jowitt made us understand the need to really look closely at the details of what was going on in dances. She helped us focus on the use of language, rhythms and the poetics of describing performance. This resulted in a qualitative shift in the way we wrote. The breadth of experience to which Banes introduced us made us aware of more sophisticated ways of thinking about ideas at play within the latest dance phenomena. We were already discussing theoretical ideas, particularly from socialist feminist scholarship which informed our writing and, in the late 1980s, Valerie suggested that Ramsay take this further by undertaking doctoral research on dance and masculinity. Concurrently, Valerie was working on the ideas that resulted in the chapter 'Dancing Dicks'. In the mid-1990s we both presented papers at the 'Border Tensions' conference at Surrey University, which not only gave us significant public exposure but also put us in touch with a broader international community of scholars who were also thinking about relations between dance theory and social and political issues.

Intellectual friendships initiated at this conference were taken further when Susan Foster helped initiate the Choreography and Corporeality Working Group under the umbrella of FIRT/IFTR. Group members' papers were circulated in advance so that when we met we could spend all our time discussing them. This process which has continued for the last decade has constantly exposed us to a range of different ideas and concerns, not only from the United States but also increasingly from continental Europe. This led to our participation in a workshop co-organised by Maaike Bleeker on corporeal literacy which explored new interdisciplinary approaches to embodiment. Foregrounding the importance of the body in thinking inspired reconfiguring relationships between art history, dramaturgy, film studies, performance theory and philosophy. Having begun our intellectual journey looking almost exclusively to the United States our focus now is largely on new European performance and new ideas coming out of Europe.

Our intellectual friendship has been based on sharing ideas and resources, reading together, introducing each other to dance works and initiating projects together. For example, Valerie introduced Ramsay to Bergson's *Matter and Memory* and Ramsay suggested Valerie looked at Agamben's essay on the irreparable. In recent years we have read together Mieke Bal's *Quoting Caravaggio* and Judith Butler's *Precarious Life* and although they hardly figure in the contents of this book they were formative to the project of writing it. Frequently, each of us, while viewing a performance, realises that it will be a useful work for the other to write about. For instance Valerie first introduced Ramsay to Monk's *Ellis Island* and Ramsay suggested Valerie see Burrows and Fargion's *Both Sitting Duet*. Over the years we have not only put together joint panels at conferences but also initiated various writing workshops, study days and symposia. Underpinning all of these activities has been sharing and circulating ideas with others and stimulating discussions about new approaches to dance works.

An extended network of fellow researchers has played an important role in our development as scholars. Within this network, reading groups have intermittently been catalysts for our writing. Together with Christy Adair and Emilyn Claid we read various writers including Deleuze and explored their relevance for dance theory. One-to-one relationships have also been important. Ruth Chandler helped Valerie appreciate the philosophical nuances and detailed workings of Deleuze's thought. Mike Huxley and Ramsay have for a long time been debating methodologies in dance history. As our focus has shifted more towards continental Europe, Martin Hargreaves was instrumental in first introducing Ramsay, in particular, to recent European performance. Isabelle Ginot, Christophe Wavelet and Claire Rousier have challenged us to reconsider our position as we've become more aware of European debates.

The basis of our writing relationship involves us regularly sending each other our work for constructive feedback knowing that the other will fully appreciate what's at stake in it. The feedback can involve 'what do you mean by this?' or suggesting ways of restructuring material. Frequently, one of us senses a thread of an argument that is not necessarily fully on the page. We often help each other make the implicit explicit. Valerie initially helped Ramsay to focus more on what the dancers were actually doing. Ramsay has often made Valerie aware that she can write more directly rather than continually feeling the need to qualify her claims. Valerie invariably draws attention to sudden jumps in Ramsay's arguments, and Ramsay in turn often challenges Valerie to clarify complex theoretical applications. Reflecting on the process of writing this book, Valerie found it not only stimulating and challenging, which she had anticipated from her experience of working with Ramsay in the past, but also enjoyable and invigorating. This is the fourth book that Ramsay has written and he has found it by far the easiest, because he has not been doing it on his own, but sharing both the dull but necessary spadework and the excitement when everything suddenly comes together.

In our process we often start with a brainstorming session. If Valerie is typing at the computer, Ramsay prowls restlessly holding forth, but not necessarily getting his own way. When Ramsay is writing on a notepad, he sometimes finds it hard to keep up with Valerie's flow of ideas. We write some pieces together from start to finish. Usually, with longer pieces, some sections are written individually and then assembled and rewritten. There have been occasions where one of us has forgotten which bit they are meant to be writing and has trespassed on the other's territory. Bits are regularly lifted from one person's section to the other's so that the ownership of sections is continually dissolving. In some older, more conservative models of scholarly practice there is a tendency for researchers to regard their areas of research as private property. The process of writing together has made us aware of differences in our ideas and concerns but, when we read what we have written, what matters is whether the ideas work. Questions about their ownership become irrelevant.

We are not suggesting that people should abandon the lone scholar model. There are, however, certain discoveries that have come out of our experience of writing together that we feel could be beneficial for the health of the community of dance scholars as a whole. A situation where people either totally agree or disagree is not a productive one. A space that allows for disagreement and criticism that is based on acceptance is productive because it stimulates change and growth. Within our process of writing together, on one level the differences between us do not matter, but on another level they are crucial because engaging with them opens up the potential for generating new insights. What we are signalling here is the importance of ethical relations with the other which involve accepting rather than fearing difference. This involves welcoming the other as equal and different, neither feeling inferior nor superior to them. This non-hierarchical approach necessitates an openness which enables an exchange and flow of energy. This is what we have been identifying in most of the dances we discuss in the book. The ongoing imperative we mentioned in the introduction has been to find new ways through our writing to combat imbalances of power. This has led us to examine ways of rethinking ethical relations with others both in our practice and in the work we have been discussing. In doing so, we think we have caught glimpses of a politics to come.

Bibliography

Acocella, J. (1992) 'Bausch's inferno'. *Art in America*, 80(1), January: 50–3.

Agamben, G. (1993) *The Coming Community*. Minneapolis: University of Minnesota Press.

Agamben, G. (2000) 'Notes on gesture', in *Means without Ends: Notes on Politics*. Minneapolis: University of Minnesota Press, pp. 49–60.

Anderson, J. (1993) 'Napoli', in M. Bremser (ed.), *International Dictionary Of Ballet – Vol. 2 : L–Z*. London and Detroit: St. James Press, pp. 996–8.

Anderson, L. (2008) personal communication, email. 6 June.

Anderson, Z. (2006) 'Yippeee!!! Sadler's Wells', *The Independent*, 8 November, [WWW] Available from: http://www.independent.co.uk/ [accessed 9.11.06].

Apter, E. (1991) *Feminizing the Fetish*. Ithaca, NY, and London: Cornell University Press.

Bal, M. (1999) *Quoting Caravaggio: Contemporary Art, Preposterous History*. Chicago and London: University of Chicago Press.

Banes, S. (1998) *Dancing Women: Female Bodies on Stage*, London and New York: Routledge.

Barthes, R. (1983) 'Inaugural Lecture, Collège de France', in S. Sontag (ed.), *Barthes: Selected Writings*. London: Fontana, pp. 457–78.

Battersby, C. (1998) *The Phenomenal Woman*. Cambridge: Polity Press.

Bauman, Z. (2003) *Liquid Love: On the Fragility of Human Bonds*. Cambridge: Polity Press.

Benjamin, W. (1973) *Illuminations*. London: Fontana.

Bentivoglio, L. (1990) 'Pina Bausch's Palermo, Palermo'. *Artforum*, April, 28(8): 19–20.

Berger, J. (1975) *A Seventh Man*. Harmondsworth: Penguin Books.

Bergson, H. (1911) *Matter and Memory*. London: Allen & Unwin.

Bergson, H. (1988) *Matter and Memory*, trans. Nancy Margaret Paul and W. Scott Palmer. New York: Zone Books.

Bershad, D. (1992) 'Looking, Power & Sexuality: Degas's '*Woman with a Lorgnette*', in R. Kendall, and G. Pollock (eds). *Dealing with Degas*. London: Pandora Press.

Bleeker, M. (2006) 'Synesthesia soirées'. *Senses & Society*, 1(3): 405–8.

Bordo, S. (1995) *Unbearable Weight*. Berkeley: University of California Press.

Boulez, P. (1995) *Boulez, Relevés d'apprenti*. Paris: Editions du Seuil.

Bournonville, A. (1979) *My Theatre Life*. London : A. and C. Black.

Briginshaw, V. (2001) *Dance, Space and Subjectivity*. Basingstoke: Palgrave Macmillan.

Brown, I. (2003a) 'The vanishing man of British dance'. *Daily Telegraph*, 13 October.

Brown, I. (2003b) 'Eloquent in their stillness'. *Daily Telegraph*, 18 October.

Buckle, R. (1975) *Nijinsky*. Harmondsworth: Penguin.

Burgin, V. (1992) Entry on 'Fantasy', in E. Wright (ed.). *Feminism & Psychoanalysis A Critical Dictionary*. Cambridge, MA: Blackwell, pp. 84–8.

Burrows, J. (2004) Unpublished email communication with the author, 3 February.
Burt, R. (1995) *The Male Dancer: Bodies, Spectacle, Sexualities*. London and New York: Routledge.
Burt, R. (2002) 'Ideologies of Englishness and internationalism in modernist ballet', in D. Peters Corbett, Ysanne Holt and Fiona Russell (eds). *The Geographies of English Art: Landscape and the National Past 1880–1940*. New Haven, CT: Yale University Press, pp. 323–45.
Burt, R. (2003) '*Both Sitting Duet*, The Place'. *Critical Dance*. [WWW] Available from http://www.criticaldance.com [accessed 8.12.03].
Butler, C. (1994) *Early Modernism: Literature, Music, and Painting in Europe, 1900–1916*. Oxford: Oxford University Press.
Butler, J. (1990) *Gender Trouble*. New York: Routledge.
Butler, J. (1993) *Bodies that Matter*. New York: Routledge.
Butler, J. (2004) *Precarious Life: The Powers of Mourning and Violence*. London: Verso.
Bynum, C. (1995) 'Why all the fuss about the body? A medievalist's perspective'. *Critical Inquiry*, 22(1), August: 1–23.
Calmette, G. (1912) 'Faux pas'. *Le Figaro*, 30 May.
Cardinal, R. (1996) 'Romantic travel', in R. Porter (ed.), *Rewriting the Self – Histories from the Renaissance to the Present*. London: Routledge.
Carter, A. (1999) 'Dying Swans or Sitting Ducks: A Critical Reflection on Feminist Gazes at Ballet'. *Performance Research*, 4(3): 91–8.
Carter, A. (2001) 'Changing views: A critical history of second wave feminist and post-feminist debate and its manifestation in writings on ballet'. *Proceedings of the Society of Dance History Scholars*, Riverside, CA: SDHS, pp. 11–15.
Carter, H. (1912) 'Ballet Bakst'. *The New Age*, 27 June, p. 211.
Chang, J. (1993) *Wild Swans*. London: Flamingo.
Cholmondeleys. (2006) 'Lea Anderson's Yippeee!!! (2006)', [WWW] available from http://www.thecholmondeleys.org/yippeee/intro.html [accessed 20.5.08].
Claid, E. (2006) *Yes? No! May be*.... New York and Abingdon: Routledge.
Code, D. (2001) 'Hearing Debussy reading Mallarmé: Music après Wagner in the Prélude à L'après-midi d'un faune'. *Journal of American Musicological Society*, 54(3), Fall: 493–556.
Cohen, P. (1980) 'Subcultural conflict and working class community', in S. Hall, D. Hobson, A. Lowe and P. Willis (eds). *Culture, Media, Language: Working Papers in Cultural Studies, 1972–79*. London : Hutchinson, pp. 78–87.
Colebrook, C. (2006) *Deleuze: A Guide for the Perplexed*. London and New York: Continuum.
Cottington, D. (1998) *Cubism in the Shadow of War: The Avant-Garde and Politics in Paris 1905–1914*. New Haven, CT, and London: Yale University Press.
Craine, D. (2006) 'Yippeee!!!', *The Times*, 7 November, T2, p. 17.
Crary, J. (1999) *Suspensions Of Perception: Attention, Spectacle, And Modern Culture*. Cambridge, MA: MIT Press.
Critchley, S. (2002) *On Humour*. London: Routledge.
Critchley, S. (2004) *Very Little...Almost Nothing* [revised 2nd edn]. London: Routledge.
Daly, A. (1987–8) 'Classical Ballet: A Discourse of Difference'. *Women & Performance*, 3(2): 57–66.

Debussy, C. (1951) *Prélude à L'après-midi d'un faune: églogue pour orchestre après Mallarmé*. London and Zurich: Ernst Eulenburg.
Deleuze, G. (1980) 'Spinoza 21/12/1980. Ontology – Ethics', [WWW] available from http://www.webdeleuze.com/pdf/uk/Spinoza/211280.zip [accessed 16/8/05].
Deleuze, G. (1989) *Cinema 2 The Time-Image*. London: The Athlone Press.
Deleuze, G. (1990) *The Logic of Sense*, trans. M. Lester with C. Stivale, ed. C. V. Boundas. New York: Columbia University Press.
Deleuze, G. (1991) *Bergsonism*, trans. H. Tomlinson and B. Habberjam. New York: Zone Books.
Deleuze, G. (1993) *The Fold*, trans. T. Conley. London: The Athlone Press.
Deleuze, G. (1994), *Difference and Repetition*, trans. Paul Patton. London: The Athlone Press.
Deleuze, G. (2003) *Francis Bacon: The Logic of Sensation*, trans. Daniel W. Smith. London and New York: Continuum.
Deleuze, G. and Guattari, F. (1988) *A Thousand Plateaus*, trans. B. Massumi. London: The Athlone Press.
Derrida, J. (2002) *On Cosmopolitanism and Forgiveness*. London: Routledge.
Dolin, A. (1969) *Pas de deux*. New York: Dover.
Dyer, R. (1985) 'Male Gay Porn Coming To Terms'. *Jump Cut*, 30, March: 27–9.
Endicott, J. A. (1999) *Je suis une femme respectable*. Paris: L'Arché [trans. Ramsay Burt, 2007].
English, R. (1980) 'Alas, Alack the representation of the ballerina'. *New Dance*, 15: 18–19.
Faludi, S. (1991/1992) *Backlash: The Undeclared War against Women*. London: Chatto & Windus.
Feldman, M. (2000) *Give My Regards to Eighth Street*. Cambridge: Exact Change.
Florence, P. (1986) *Mallarmé, Monet and Redon: Visual and Aural Signs and the Generation of Meaning*. Cambridge: Cambridge University Press.
Flusser, V. (2002) 'Taking up residence in homelessness', in *Writings*. Minneapolis: University of Minnesota Press, pp. 91–103.
Flusser, V. (2003) 'The challenge of the migrant', in *The Freedom of the Migrant*. Champaign, IL: University of Illinois University Press, pp. 1–15.
Forrester, J. (1987) 'A brief history of the subject', in L. Appignanesi (ed.), *ICA Documents 6, Identity*. London: ICA, pp. 12–16.
Foster, S. L. (1996) 'The Ballerina's Phallic Pointe', in *Corporealities*. New York: Routledge.
Foucault, M. (1979) *The History of Sexuality Volume 1: An Introduction*. Harmondsworth: Allen Lane.
Franko, M. (2002) *The Work of Dance: Labor, Movement and Identity in the 1930s*. Middletown, CT.: Wesleyan University Press.
Freud, S. (1962) *Three Essays on the Theory of Sexuality*, trans. James Strachey. New York: Basic Books.
Garafola, L. (1989) *Diaghilev's Ballets Russes*. New York and Oxford: Oxford University Press.
Gatens, M. (2000) 'Feminism as "Password": Re-thinking the "Possible" with Spinoza and Deleuze'. *Hypatia*, Spring 15(2): 59–75.
Gautier, T. (1932) *The Romantic Ballet as seen by Theophile Gautier*, trans. C. W. Beaumont. London: C. W. Beaumont.

Gil, J. (2006) 'Paradoxical Body'. *TDR*, 50(4), Winter: 21–35.
Gilroy, P. (1995) '"...to be real": The dissident forms of black expressive culture', in C. Ugwu (ed.), *Let's Get It On: The Politics of Black Performance*. London: ICA, Seattle: Bay Press.
Ginot, I. (2007) 'Identity, the contemporary, and the dancers', in Susanne Franco and Marina Nordiera (eds), *Dance Discourses: Keywords in Dance Research*. London and New York: Routledge, pp. 251–66.
Gold, A. and Fizdale, R. (1979) *Misia: The Life of Misia Sert*. New York: Alfred A Knopf.
Goldberg, M. (1991) 'Trisha Brown's Accumulations'. *Dance Theatre Journal*, 9(2), Autumn: 4–7, 39.
Gottschild, B. D. (1996)) *Digging the Africanist Presence in American Performance: Dance and Other Contexts*. Westport, CT: Greenwood Press.
Greskovic, R. (1985) 'Armitagean Physics, or The Shoes of the Ballerina'. *Ballet Review*, 13(2), Summer: 73–89.
Grosz, E. (1992) Entry on 'Voyeurism/Exhibitionism/The Gaze', in E. Wright (ed.), *Feminism & Psychoanalysis A Critical Dictionary*. Cambridge, MA: Blackwell, pp. 447–50.
Grosz, E. (1994) *Volatile Bodies*. Bloomington and Indianapolis: Indiana University Press.
Guest, A. H. and Jeschke, C. (1991) *Nijinsky's Faune Restored: A Study of Vaslav Nijinsky's 1915 Dance Score, L'après-midi d'un faune, and His Dance Notation System Revealed*. Philadelphia: Gordon & Breach.
Guest, I. (1972) *The Romantic Ballet in England*. London: Pitman.
Hargreaves, M. (1998) 'The pelvis unbound: An investigation into the disruptive nature of Elvis Presley's dancing'. *Theater Insight*, Spring, 8(2): 3–9.
Hargreaves, M. (2007) 'Diseased Dames in Dancing Shoes'. *Dance Theatre Journal*, 22(2): 23–6.
Haskell, A. (1989) 'Introduction', in A. Dolin, *Pas de deux*. New York: Dover.
Hebdidge, D. (1979) *Subculture: The Meaning of Style*. London: Methuen.
Howes, F. (1965) *The Music of William Walton*. Oxford: Oxford University Press.
Hutera, D. (2003) 'Both Talking' extracts from a post-show interview with Jonathan Burrows and Matteo Fargion'. *Dance Umbrella News*, Autumn.
Järvinen, H. (2003) *The Myth and Genius in Movement: Historical Deconstruction of the Nijinsky Legend*. Unpublished thesis (PhD), Turko: Turun Yliopisto.
Järvinen, H. (2007) 'Some steps towards a historical epistemology of corporeality', in *Proceedings of the 30th Annual Conference of the Society for Dance History Scholars*, Paris, 21–24 June 2007: 145–8.
Jeitschko, M. (1989) 'Pina Bausch – Berlin is her Palermo'. *Welt am Sontag*, 31 December. n.p.
Jordan, S. and Grau, A. eds (1996) *Following Sir Fred's Steps*. London: Dance Books.
Julien, A. (1912) 'Revue musicale'. *Journal des Débats*, 9 June: 2.
Kane, A. and Pritchard, J. (1994) 'The Camargo Society Part 1'. *Dance Research*, XII(2), Autumn: 63–4.
Kant, I. (1983) *Perpetual Peace, and Other Essays* [1795]. Indianapolis: Hackett.
Kavanagh, J. (1996) *Secret Muses: The Life of Frederick Ashton*. London: Faber & Faber.

Kayser, L. (1990) 'Carnet culturel'. *D'Letzburger Land*, 12 January: 17.

Koegler, H. (1990) 'Pina Bausch/Tanztheater Wuppertal'. *Dance Magazine*, April: 74–5.

Kracauer, S. (1995) 'Travel and dance', in *The Mass Ornament*, ed., trans. and with an introduction by Thomas Y. Levin. Cambridge, MA: Harvard University Press, pp. 65–74.

Kristeva, J. (1991) *Strangers to Ourselves*. New York: Columbia University Press.

Kristeva, J. (1993) *Nations Without Nationalism*. New York: Columbia University Press.

Kudláček, M. (2007) 'Isaac Julien'. *Bomb*, 101, Fall: 72–9.

Lambert, C. (2002) 'On being moved: Rainer and the aesthetics of empathy', in *Yvonne Rainer: Radical Juxtapositions 1961–2002*. Philadelphia: University of the Arts Press and Rosenwald-Wolf Gallery, pp. 116–31.

Laqueur, T. W., (1995) 'Credit, novels, masturbation, in Susan Foster (ed.) *Choreographing History*. Bloomington and Indianapolis: Indianapolis University Press, pp. 119–28.

Lavin, M. (1993) *Cut With the Kitchen Knife*. New Haven, CT, and London: Yale University Press.

Le Roy, X. (1998) *Product of Circumstances*, [WWW] available from: http://www.unfriendly-takeover.de/f14_Videos.htm [accessed 21/10/07].

Le Roy, X. (2000) 'Self-Interview', [WWW] available from: http://www.insituproductions.net/ [accessed 1/7/05].

Lesure, F. and Nichols, R. (1987) *Debussy's Letters*. London and Boston: Faber & Faber.

Lloyd, G. (1996) *Spinoza and the Ethics*. London and New York: Routledge.

Mac Low, C. (2006) *'AG Indexical, with a little help from H.M.* Yvonne Rainer talking with Clarinda Mac Low interview / 4.3.06', [WWW] available from: http://movementresearch.org/publishing/cc/interview/rainer1.html/ [accessed 25/4/06].

MacCannell, J. F. (1992) Entry on 'Desire', in E. Wright (ed.), *Feminism & Psychoanalysis: A Critical Dictionary*. Cambridge, MA: Blackwell, pp. 63–8.

Macey, D. (1992) Entry on 'Phallus: definitions', in E. Wright (ed.), *Feminism & Psychoanalysis: A Critical Dictionary*. Cambridge, MA: Blackwell, pp. 318–20.

Mackrell, J. (2006) 'Yippeee!!!, Sadler's Wells, London'. *The Guardian*, 6 November, p. 36.

Mackrell, J. (2008) *The Bloomsbury Ballerina: Lydia Lopokova, Imperial Dancer And Mrs John Maynard Keynes*. London: Weidenfeld & Nicolson.

Mallarmé, S. (1976) *Igitur, Divagations, Un Coup de dés*. Paris: Editions Gallimard.

Manning, E. (2007) *Politics of Touch*. Minneapolis: Minnesota University Press.

Manning, E. (2008) 'Introduction to *Relationscapes: Movement, Art,* Philosophy'. Cambridge, MA: MIT Press (in press). [WWW] Available from: http://www.erinmanning.lunarpages.net [accessed 20.5.08].

Manning, S. (2006a) 'Letter from the President'. *Society of Dance History Scholars Newsletter*, XXV1(1), Spring: 1–2.

Manning, S. (2006b) 'Letter from the President'. *Society of Dance History Scholars Newsletter*, XXV1(2), Fall: 1–2.

Maranca, B. (1980) 'Meredith Monk's Recent Ruins: The archaeology of consciousness, essaying images'. *Performing Arts Journal* IV(3): 12, 39–49.

Massumi, B. (2002) *Parables for the Virtual: Movement, Affect, Sensation*. Durham, NC, and London: Duke University Press.
McAndrew, P. (2002) 'Darling Helene: August Bournonville's letters from France and Italy, 1841 part three'. *Dance Chronicle*, Winter 25(3): 343–445.
McCarren, F. (1995) 'The "symptomatic act" circa 1900: Hysteria, hypnosis, electricity, dance'. *Critical Inquiry*, 21 (Summer): 748–74.
McClary, S. (1991) *Feminine Endings: Music, Gender and Sexuality*. Minneapolis: University of Minneapolis Press.
McMahon, B. (2005) 'Asylum seekers' voyages of hell'. *The Observer*, 17 July, n.p.
McQuirter, M. (2002) 'Awkward moves: Dance lessons in the 1940s', in T. DeFrantz (ed.), *Dancing Many Drums*. Madison, WI: Wisconsin University Press, pp. 81–104.
Meisner, N. (1992) 'Come dance with me'. *Dance and Dancers*, November: 12–16.
Mellers, W. and Hildyard, R. (1995) 'The Cultural and Social Setting', in Boris Ford (ed.), *The Cambridge Cultural History: Volume 8 Early Twentieth Century*. Cambridge: Cambridge University Press, pp. 2–24.
Money, K. (1992) 'Margot – A Photographer's Story'. *Daily Telegraph Magazine* January.
Mulvey, L. (1973) 'You Don't Know What is Happening, Do You Mr. Jones?'. *Spare Rib*, 8: 13–16, 30.
Mulvey, L. (1975) 'Visual Pleasure and Narrative Cinema'. *Screen*, 16(3), Autumn: 6–18.
Nash, M. (2008) 'Reality in the Age of Aesthetics'. *Frieze*, 114, April, [WWW] available from: http://www.frieze.com/magazine/ [accessed 17/4/08].
Newman, B. (1987) 'Lea Anderson of The Cholmondeleys'. *Dancing Times*, November: 130–2.
Nijinska, B. (1981) *Early Memoirs*. London and Boston: Faber & Faber.
Nijinska, B. (1986) 'On movement and the school of movement', in Nancy Van Norman Beer (ed.), *Bronislava Nijinska: A Dancer's Legacy*. San Francisco: The Fine Arts Museum of San Francisco, pp. 85–7.
Nijinsky, R. (1970) *Nijinsky*. London: Sphere Books.
Nijinsky, V. (1912) 'Les souvenirs de Nijinsky et Karsavina par eux-mêmes'. *Je sais tout*, 15 November: 406–20.
Pearson, M. and Shanks, M. (2001) *Theatre/Archaeology*. London: Routledge.
Petridis, A. (2004) 'Will the creator of modern music please stand up?'. *The Guardian*, 16 April, G2, p. 4.
Phelan, P. (1993) *Unmarked*. New York: Routledge.
Phelan, Peggy. (1999) 'Yvonne Rainer: From dance to film', in Y. Rainer (ed.), *A Woman Who…Essays, Interviews, Scripts*. Baltimore, MD, and London: Johns Hopkins University Press.
Provvedini, C. (1990) 'Pina Bausch: parla la coreografa tedesca al debutto con il suo spettacolo al Teatro Biondo', trans. Diana Crampton. *Italia Oggi*, 22 January: 2–3.
Quirey, B. (1976) *May I Have The Pleasure? The Story Of Popular Dance*, London: BBC.
Rainer, Y. (1974) *Work 1961–73*. Halifax, NS: The Press of Novia Scotia College of Art and Design.
Rambert, M. (1972) *Quicksilver*. Basingstoke: Macmillan.

Redon, O. (1912) letter, *Le Figaro,* 31 May.
Regitz, H. (1998) '"I'm far too self-critical": An interview with Josephine Anne Endicott'. *Ballett International/Tanz Aktuel,* 8–9: 50–5.
Rivière, J. (1912) 'L'après-midi d'un faune de Nijinski', *Excelsior,* 28 May: n.p.
Roach, J. (1985) *The Player's Passion: Studies in the Science of Acting.* Delaware: University of Delaware Press.
Rockwell, J. (2006) 'Connecting the Dots Between Stravinsky and Hip-Hop'. *New York Times,* 21 April, [WWW] available from: http://www.nytimes.com/2006/04/21/arts/dance/15stra.html [accessed 25/4/06].
Sachs, O. (1985) *The Man Who Mistook His Wife for a Hat.* London: Duckworth.
Salle, D. (2006–7) 'David Salle on Karole Armitage'. *Modern Painters,* December 2006– January 2007: 9–10.
Savigliano, M. (1998) 'From Wallflowers to femmes fatales: Tango and the performance of passionate femininity', in William Washabaugh (ed.) *The Passion of Music and Dance : Body, Gender and Sexuality.* Oxford : Berg, pp. 103–10.
Sayers, D. L. (1977) *Have His Carcase.* London: New English Library.
Schneider, L. (1912) 'Les Ballets Russes'. *Le Théâtre,* 11 May: 6–9.
Schor, N. (1992) Entry on 'Fetishism', in E. Wright (ed.), *Feminism & Psychoanalysis: A Critical Dictionary.* Cambridge, MA: Blackwell, pp.113–17.
Silverman, K. (1988) *The Acoustic Mirror.* Bloomington and Indianapolis: Indiana University Press.
Sitwell, E. (1987) *Façade: With an Interpretation by Pamela Hunter.* London: Duckworth.
Skene-Wenzel, J. (2003) 'Both Sitting Duet'. *Londondance,* [WWW] available from http://www.londondance.com [accessed 8.12.03].
Sokolova, L. (1960) *Dancing for Diaghilev.* London: John Murray.
Spinoza, B. (2001) *Ethics.* Ware: Wordsworth Editions.
Tenroc, C. (1912) 'Nijinsky va faire dans *L'Après-midi d'un faune* des essays de chorégraphie cubiste'. *Comoedia,* 18 April: 4.
Touchard, M. (1912) 'Les Ballets Russes'. *La Nouvelle Revue,* 15 July: 201–10.
Vuillermos, E. (1912) 'La grande saison de Paris'. *S.I.M. Révue Musicale,* June: 62–8.
Williams, A. (2003) 'Jonathan Burrows and Matteo Fargion "Both Sitting Duet"'. *Ballet Magazine,* December, [WWW] available from http://www.ballet.co.uk [accessed 8.12.03].
Williams, R. (1989) *The Politics Of Modernism.* London: Verso.
Wolff, J. (1990) *Feminine Sentences.* Cambridge: Polity Press.
Wolff, J. (1996) 'The failure of the hard sponge: Class, ethnicity, and the art of Mark Gertler'. *New Formations,* 28, Spring: 56.
Wright, E. ed. (1992) *Feminism & Psychoanalysis: A Critical Dictionary.* Cambridge, MA: Blackwell.
Zimmer, E. (1986) 'Out there with Karole Armitage'. *Dance Magazine,* May: 38–41.

Index